elementary geometry

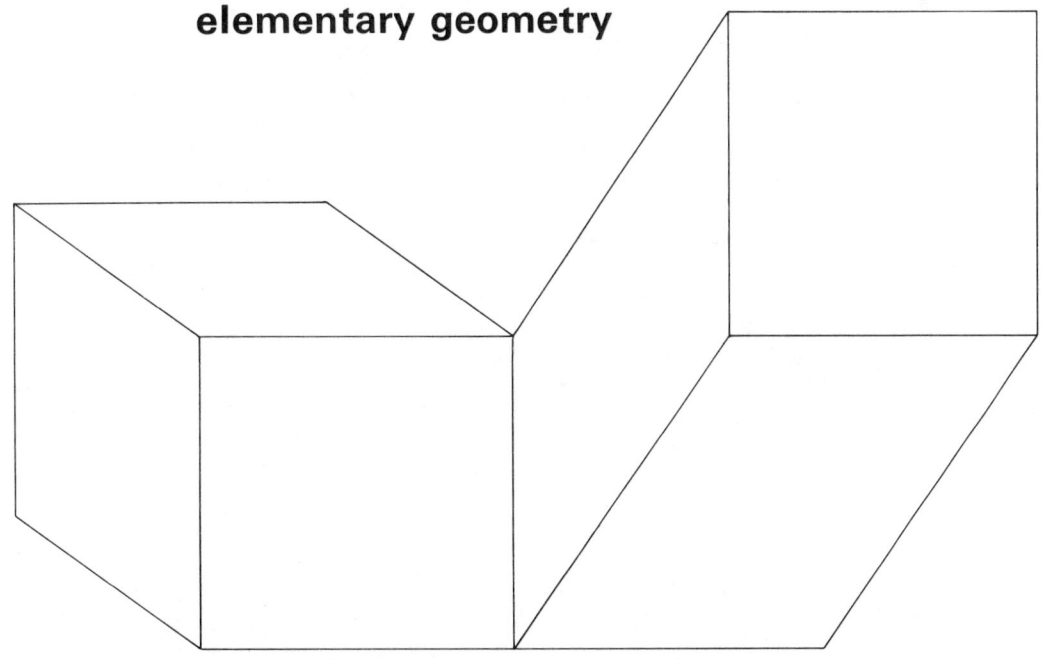

Vincent H. Haag
Franklin and Marshall College

Clarence E. Hardgrove
Northern Illinois University

Shirley A. Hill
University of Missouri—Kansas City

Addison-Wesley Publishing Company
Reading, Massachusetts
Menlo Park, California
London · Don Mills, Ontario

For example, the edge of a book, a stretched string, and a ray of light all have a property that the edge of a coin, a slack string held at each end, and a wire made into a coil do not have. These observations form a basis for the idea of straightness, which is used to describe some sticks and the edges of some tables.

Our school experiences help us to clarify ideas as we distinguish between the objects (the edge of the book, the string, and the ray of light) and their common property of straightness. We learn to separate the two, the object and the idea. As we do this, a concept of a geometric figure—a straight line—emerges. This straight line is a conceptual model of all straight edges on physical objects. We use a stretched string or a picture of a stretched string as a physical model of a straight line, but as we talk about a straight line we distinguish between the object and the concept.

In this same way, a conceptual model of a physical position is conceived and called a point, and a conceptual model of all flat surfaces is conceived and called a plane. Later, when we begin to think of a model of space as a set of points, we think of lines and planes as special sets of points in this space.

The development of geometry in this way lacks the formality of the next stage. We make observations in our physical world; we conceive ideas and clarify and expand them; we relate them and classify them; and we understand that they apply generally to classes of physical objects. The geometry developed in this way is called *informal geometry*.

1–3
points, lines, and planes

points The notion of a point results from an observation of reoccurring features in the physical world. We describe objects such as pins, boxes, pencils, and books by identifying their common properties. A small tip or sharp corner is a feature which we can see and feel and which is important to the description of each of these objects. Thus the tip of a child's pencil shows a position in the room, and the tip at the corner of a book and a small dot on a girl's dress show other positions. These tips are at first called points, as are also dots drawn on the blackboard to show positions. In a classroom, a dot on a paper can be accepted as a picture of

4   informal geometry

a point, and a letter written beside the dot is a name for the point. The point itself is not the dot but what we imagine when we look at the dot.

lines   Points locate positions. We can feel some points (pinpoints) with our fingers and see other points (such as steeple tips) with our eyes. We can also feel edges of objects and see them, but we cannot easily test the straightness of an edge by touch. It is, however, easy to sight along an edge to decide whether it is straight, or to stretch a piece of string along an edge. If every part of the edge is on the line of sight or is on the stretched string, the edge has straightness. The corresponding mental image of the straight edge or of the piece of stretched string is a *line segment*. A straight edge has two endpoints, and so too does a line segment. When we sight along edges between their two endpoints or stretch a string from point to point, it is intuitively evident that exactly one line segment is determined by any two endpoints.

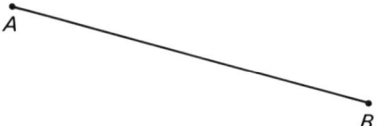

The line segment with endpoints A and B is denoted by $\overline{AB}$. If we sight along $\overline{AB}$ from A toward B, the line of sight ideally continues without end. For example, let A represent the eyepiece of a telescope, B represent the other end of the scope that points the line of sight to a star. The geometric concept that results from experiences like this is called a *ray*. The ray from A through B is denoted by $\overrightarrow{AB}$ and read "ray AB." The ray from B through A is named $\overrightarrow{BA}$ and read "ray BA." The drawing below shows $\overrightarrow{AB}$.

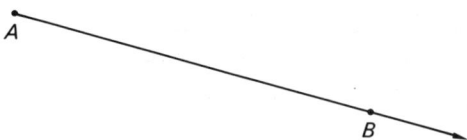

If the line of sight is extended in both directions, the resulting concept is called a *line*. The line through A and B is named $\overleftrightarrow{AB}$ and read "line AB." The drawing on page 5 shows $\overleftrightarrow{AB}$.

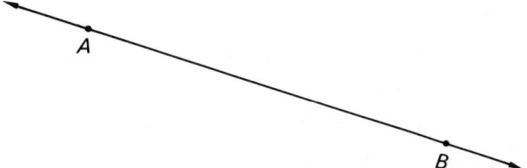

There are no experiences in real life that directly support the concepts of a ray or a line; no part of a physical object that can be felt or seen shows a ray or a line. This extension of the idea of a line segment to the ideas of a ray and a line is the first opportunity a child has to use his geometric imagination to the fullest.

There are distinctions to be made between parts of objects, pictures of the parts, and the geometric ideas which the parts suggest. For example, we can distinguish among the tips of corners of objects, dots drawn on a blackboard, and points; and we can distinguish among stretched strings, pictures of line segments drawn on paper, and line segments. A child does not make these distinctions as he begins to think about geometry, but he makes them when he is ready. As the ideas develop, children should be allowed to use any language that they believe is descriptive. Teachers, however, should use language that supports the distinction between the levels of development of an idea. This language helps the children ultimately to make the distinction.

planes The examination and description of objects that are "flat" lead to an intuitive idea of a plane. As we describe the one surface of a ball, the three surfaces of a tin can, and the six surfaces of an ordinary box, the idea of flatness emerges. "A surface is flat if it is full of straightness," that is, if it is straight along every line of sight. For example, the surface of a water pipe is not flat; it is straight along some lines of sight, but it is not straight along other lines of sight. A table top is flat because it is straight along every line of sight. The mental image which corresponds to all flat surfaces is the *plane*. Again the imagination of the child is called upon to think of a plane as extending without end in all directions and being straight along all lines of sight.

properties relating points, lines, and planes As children draw and use models of points, lines, and planes they discover relationships among them. For example, with dots as models for points, wires for lines, and cardboards for planes, children can experiment and decide that:

> Given two points, there is exactly one line through the points; and there is exactly one line segment which has the two endpoints.
>
> Given two lines, there is at most one point common to the two lines (or the two lines intersect in one point or do not intersect at all).
>
> If two lines have a point in common (if the lines intersect), then there is exactly one plane containing both lines.
>
> Given three points (not all on the same line), there is exactly one plane through the three points.
>
> Given two planes, they either intersect in a line or they do not intersect.

language As children discover relationships among points, lines, and planes, they need language to talk about their discoveries. Any language that is natural to the children and conveys the meaning they intend is acceptable. For example,

"The point is on the line" has the same meaning as "The line contains the point" or "The line passes through the point."

"The point is on the plane" has the same meaning as "The plane contains the point" or "The plane passes through the point."

"The line is on the plane" has the same meaning as "The plane contains the line" or "The plane passes through the line."

The geometric language a person uses depends on his feeling about geometric figures as being either static or dynamic in origin. The statement, "The line contains the point" suggests a line as a fixed entity made up of points. This viewpoint, which considers figures as sets of points, is discussed in Chapter 3. The statement, "The line passes through the point" suggests a line as a path of a moving point. This dynamic approach in which movements of figures suggest some properties of the figures is discussed later in this chapter, and will be investigated further in Chapters 4 and 6. Both the static and dynamic views lead to fruitful results; there is no contradiction between the views. We need all ways available to us when we are experimenting and discovering ideas.

1-4
plane and space figures

Our imagination and intuition lead us to quickly classify plane figures in the usual categories such as line segments, triangles, and quadrilaterals.

When we start with two distinct points, say A and B, there is exactly one *line segment* $\overline{AB}$ joining them. But when we are given a third point not on $\overleftrightarrow{AB}$, say C, we may draw exactly one three-sided figure, called a *triangle* (Fig. 1–1). It consists of three line segments with endpoints at A or B or C. The language "the point is on the triangle" means "the point is on one of the three line segments forming the triangle."

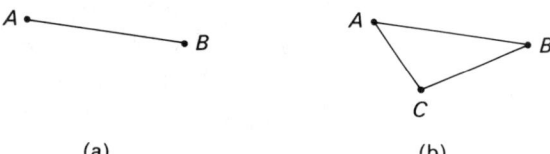

Fig. 1–1. (a) Line segment determined by A and B. (b) Triangle determined by A, B, and C.

(a)   (b)

Given a fourth point, say D, in the plane of A, B, and C but not on triangle ABC, we may draw several four-sided figures, called *quadrilaterals* (Fig. 1–2). Quadrilaterals are those figures which consist of four line segments that have endpoints among the points A, B, C, and D, and such that the segments intersect only in endpoints.

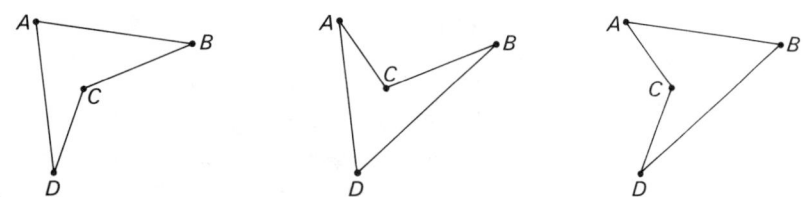

Quadrilaterals determined by A, B, C and D

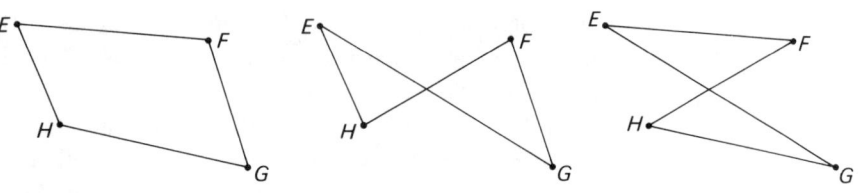

Figure 1–2   Quadrilateral *EFGH*   Not quadrilaterals

Similarly, given a fifth point, we can show a five-sided figure (pentagon), and given a sixth point we can show a six-sided figure, etc. Triangles, quadrilaterals, and other figures of this type, consisting of line segments that intersect only in endpoints, are called *polygons*.

congruent line segments Plane figures can be classified by size as well as by shape. The line segments $\overline{AB}$ and $\overline{CD}$ shown below are the same shape (all segments are the same shape), but we sense a difference in their size. If we desire to test this possibility, we can choose two objects that suggest the two line segments, such as two pieces of wire. We hold one piece that suggests $\overline{AB}$ against the other piece that suggests $\overline{CD}$ and decide whether the endpoints of one fit exactly on the endpoints of the other. If the endpoints fit, we say that the corresponding line segments are the same size. If not, they are of different sizes.

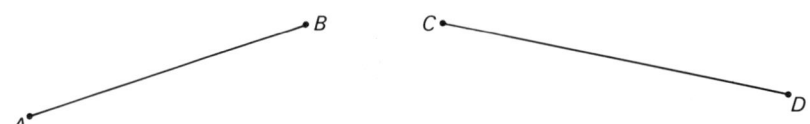

Again language must be considered. When we say, "$\overline{AB}$ and $\overline{CD}$ are equal," written $\overline{AB} = \overline{CD}$, we mean that "$\overline{AB}$" and "$\overline{CD}$" are names for the same line segment. Thus we cannot say that two different segments of the same size are equal. The two segments are not equal because they are different line segments. We say that the line segments are *congruent*. The word congruent means "coinciding" in the sense that one segment can be "moved" to coincide with the other. Here the intuitive notion of the movement of the figure is used to mean a movement of an object which suggests the figure. Movements in space are discussed in Section 1-5 of this chapter; in Chapter 4, we make the concept of the motion of a figure more precise because it is, in fact, a key concept in geometry.

In a similar way, two triangles, two quadrilaterals, and other plane figures can be compared by size as well as by shape. If one figure can be moved so as to coincide with the other, that is, if each line segment of the one coincides with the corresponding line segment of the other, the figures are called *congruent* figures.

angles Our intuition about angles is possibly weaker than our intuition about lines and figures consisting of line segments. We can stimulate intuition, however, as we show familiar examples of corners on objects, such as a corner of a picture frame, and discuss them. Tracing a corner with two fingers suggests that a corner is formed by two line segments with one common endpoint. From this beginning notion of corner, we can again call on imagination to extend the two line segments to rays. With this extension we form the intuitive idea of an *angle* as a pair of rays with a common endpoint (Fig. 1–3). (At first we exclude the case in which the rays are on the same line.) If the angle is formed by the rays $\overrightarrow{AB}$ and $\overrightarrow{AC}$, we denote the angle by $\angle BAC$ or $\angle CAB$. Point A is the common endpoint of the two rays and is called the *vertex* of the angle. The letter A is the middle letter of the symbol $\angle BAC$ to signify that the point A is the vertex of the angle.

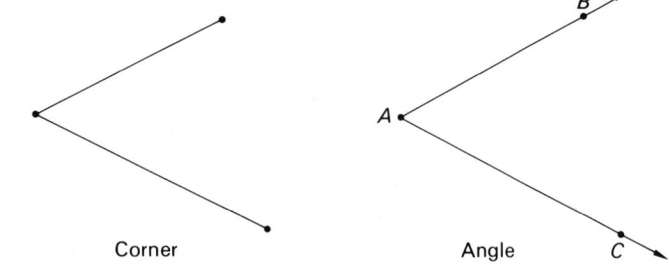

Figure 1–3   Corner   Angle

congruent angles Each angle consists of a pair of rays with a common endpoint. Not all angles are the same size. The question of whether two given angles are the same size can be determined by a simple experiment. Paper can be folded to make a model of one angle. One fold of a piece of paper produces a straight edge; another fold produces a corner to fit one of the angles. In Fig. 1–4 a paper is folded to fit $\angle ABC$. This paper corner is then placed on $\angle MPT$ to test whether a copy of $\angle ABC$ fits $\angle MPT$, that is, whether the angles are congruent.

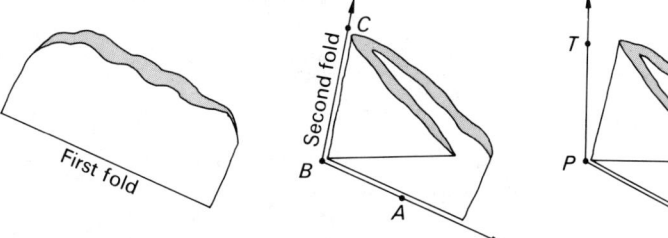

Fig. 1–4. The angles are not congruent.

10   informal geometry

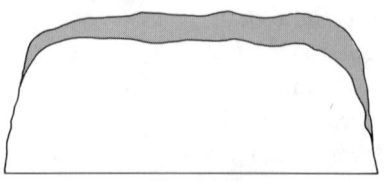
First fold

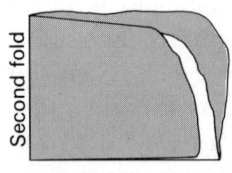

Fig. 1–5. Making a copy of a square corner.

Among the corners of objects, the square corner is the most common. By experimenting with paper-folding, we can make a copy of a square corner (Fig. 1–5). The angle shown by a square corner is called a *right angle*. It has properties that can easily be discovered, such as the property that if one and only one ray of one right angle coincides with one and only one ray of another right angle, then their other rays form a line.

plane curves and plane regions We can draw pictures of certain plane figures, such as quadrilaterals, on a plane surface by moving a pencil from one point to another without lifting the pencil from the surface. Any plane figure whose picture can be drawn in this way is called a *plane path*. Note that a plane path has a beginning point and an ending point. The figure suggested by a plane path is usually called a *plane curve*. For example, if a piece of string is dropped on a flat surface, it shows or suggests a plane curve.

If the drawing of the plane path ends at the same point at which it begins, the path is called *closed*. The corresponding closed curve can be shown by a string with its ends tied together. Most interesting closed curves in geometry, such as a triangle and a quadrilateral, are simple closed curves (Fig. 1–6). A *simple* closed curve does not intersect itself.

A curve, not closed

A closed curve, not simple

Plane figure

A simple closed curve

Figure 1–6

Simple closed curves have special importance in geometry because they are boundaries of plane regions. For example, a

Figure 1–7 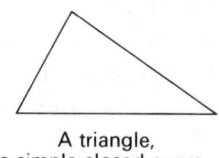 A triangle, a simple closed curve 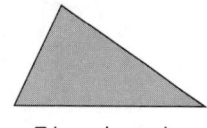 Triangular region

triangle (Fig. 1–7) is a simple closed plane curve; the part of the plane inside the triangle, along with the triangle itself, is a triangular region. The triangle is the boundary of this region. In the same way, every simple closed curve in a plane has an *interior*, the part of the plane inside the curve. The curve together with its interior is a *plane region*.

Figure 1–8  Rectangle  Rectangular region 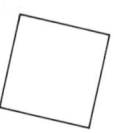 Square  Square region

As children identify and name figures, both curves and regions, they naturally classify them. For example, as they learn about rectangles and squares and the regions formed by these figures (Fig. 1–8) their descriptions—which may be similar to the ones below—show this classification.

A rectangle is a quadrilateral with all its corners square.
A square is a rectangle with all its sides congruent.

Children may have classified a circle intuitively as a simple closed plane curve with "roundness." (Section 1–5 discusses circles further.)

simplex In advanced mathematics a point is called a 0-*simplex* (Fig. 1–9); two points with the segment joining them is called a 1-*simplex;* a point, a line segment (not on the point), together with all the line segments joining the given point to points of the line segment is called a 2-*simplex*. Note that a 2-simplex is a triangular region; each side is a 1-simplex, and each vertex is a 0-simplex.

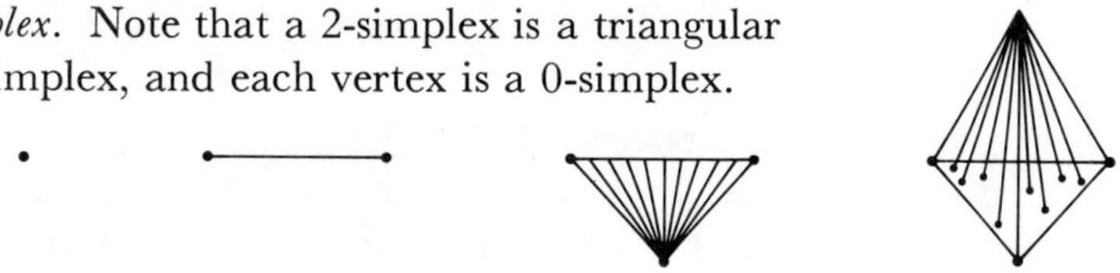

Figure 1–9   0-simplex   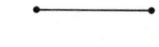 1-simplex   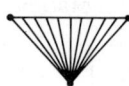 2-simplex   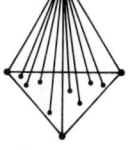 3-simplex

The natural extension is to define a 3-simplex as a figure formed by a triangular region (a 2-simplex), a point not on the plane of the triangle, and all the segments joining the given point to points of the triangular region. This is an example of a *space figure*.

figures in space Physical objects occupy space and do not themselves suggest plane figures. It is the surfaces of the objects that suggest plane figures. Simple space figures may be classified in terms of their surfaces:

a) Figures whose surfaces are all plane regions
b) Figures with some surfaces that are plane regions and others that are "rounded" regions
c) Figures whose surfaces are all "rounded" regions

An object shaped like a brick suggests a space figure of the first type. Its surfaces consist of six rectangular regions, called its *faces*. The faces intersect in pairs to form twelve line segments called *edges*. The eight endpoints of the edges are called *vertices*. The surface of the brick is of the form of a box, and is technically called a *rectangular prism*, but the word *box* is more descriptive and appropriate language for children who are beginning a geometry program. Another space figure of this type is the 3-simplex described and shown above; its surface is called a *triangular pyramid* or *tetrahedron*. Its four faces are triangular regions. The plane faces intersect in six edges, and the endpoints of the edges form four vertices.

An example of type (b) space figure is suggested by a closed tin can. The space figure suggested by the can is called a *cylinder*. It has two surfaces that are plane circular regions and another surface that is rounded.

A solid ball suggests a space figure of type (c). Its only surface is completely rounded; the surface of a ball is called a *sphere*.

exercise set 1-1

1. This statement is false: Given any four points in space, then the four points in space are vertices of a quadrilateral. Why?

2. a) Locate five points in the plane of a paper, no three of which are on the same line. Show all the 5-sided figures that have these points as vertices.
   b) Which of these figures are 5-sided polygons (pentagons)?

3. Which of these sentences are false? Make them true by the deletion or change in one and only one word.
   a) A triangle is a simple closed curve.
   b) The symbol for 8 made by a typewriter is a simple closed curve.
   c) The symbol for p made by a typewriter is a closed curve.

4. Which of these statements are true? Explain.
   a) $\overline{AB} = \overline{BA}$
   b) $\overleftrightarrow{AB} = \overleftrightarrow{BA}$
   c) $\overrightarrow{AB} = \overrightarrow{BA}$
   d) $\angle BAC = \angle CAB$
   e) $\angle ABC = \angle ACB$

5. Upon what property given on page 6 does the following statement depend? Explain your answer.
   Given a line and a point not on the line, there is exactly one plane through the line and the point.

6. Which of the figures in Fig. 1–10 are curves? Which are simple closed curves? Which are closed curves?

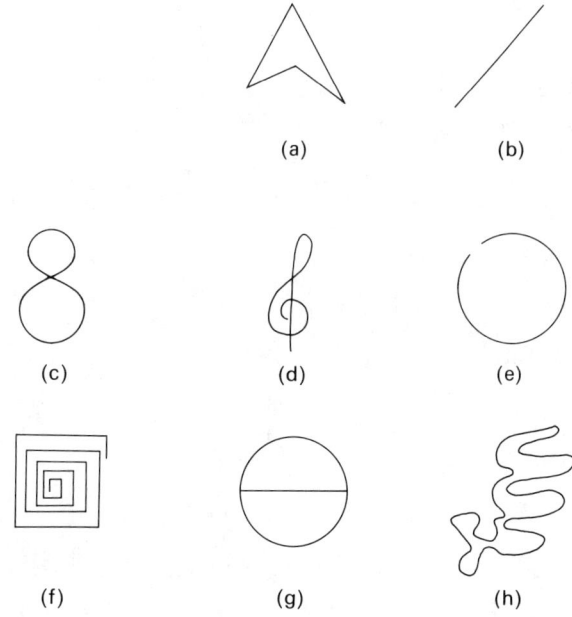

Figure 1–10

Space Figures	Number of faces, f	Number of vertices, v	f + v	Number of edges, e	Comparison of (f + v) and e
Triangular prism					
Rectangular prism					
Rectangular pyramid					
Tetrahedron					
Octagonal pyramid					
Octagonal prism					
Pentagonal pyramid					
Pentagonal prism					

7. a) Complete this chart.

   b) Make a conjecture about the relation of $(f + v)$ and e of these space figures.

8. What geometric figure is contained in both $\overrightarrow{BA}$ and $\overrightarrow{AB}$?

9. Name all the angles in each part of Fig. 1–11.

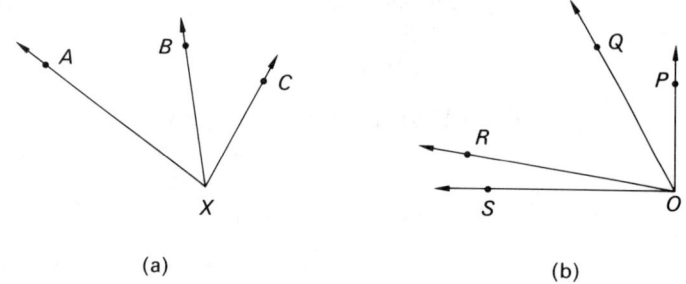

Figure 1–11   (a)   (b)

10. For the rectangular prism and rectangular pyramid shown in Fig. 1–12, describe, if possible:
   a) A pair of faces contained in parallel planes, that is, planes with no point in common
   b) A pair of edges contained in parallel lines
   c) A pair of edges contained in intersecting lines
   d) A pair of edges contained in lines that do not intersect and are not parallel
   e) Three faces contained in planes that have a point in common
   f) Four faces contained in planes that have a point in common
   g) An edge contained in a line which is the intersection of two planes
   h) A point that is the vertex of a right angle

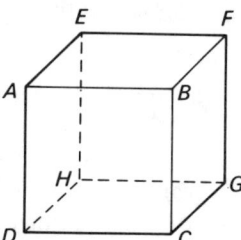
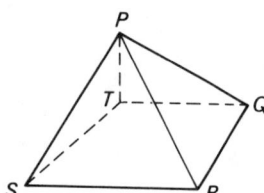

Figure 1–12

teaching questions and projects 1–1

1. a) Describe the notion you may have had of a circle at about age 5; at about age 10; at about age 16.
   b) Talk with a person of one of these ages and try to determine his idea of a circle.

2. Name three physical objects that are models of: (a) a point, (b) straightness, (c) a straight edge, (d) flatness, (e) a surface.

3. Describe an activity or present an argument that helps a child to distinguish between a physical model of a point and the idea of a point.

4. a) What imaginary geometric model should a child form for all the straight edges he finds in his physical environment?
   b) For all the flat surfaces?
   c) For all the tips of corners?

5. a) What geometric concepts do these models help develop?
      i) A taut rubber band that is held tightly at each end.
      ii) An ever-expanding elastic thread stretched between the beaks of two magic birds who fly forever farther and farther apart.
      iii) An ever-expanding elastic thread stretched from your hand to the beak of a magic bird who flies forever farther and farther from you.
   b) Give a description of a corresponding model that would help to develop the concept of a plane.

6. a) Describe how you can guide a child to discover this property: If a line and a plane intersect, the intersection is a point on the line itself. (What physical objects will aid the child? What questions can you ask to guide discovery?)
   b) How is a child likely to state the property after he discovers it?

7. Repeat Exercise 6 for one of the other properties given in Section 1–3, page 6.

8. Which of these statements are false?
   a) A taut string is a more suitable model of a line segment than an edge formed by the fold of a paper.
   b) A triangular cardboard cutout is a more suitable model of a triangle than a triangle formed by bending a piece of fine wire.
   c) A metal hoop is a more suitable model of a circle than a cardboard disc.
   d) A fine wire bent in one place is a more suitable model of an angle than the corner of this page.

9. Suppose a child says, "A circle is a square with its corners cut off." How might a teacher react to this language?

10. Imagine that you drop a piece of string on the floor many times. Which of these figures might possibly be shown at one or more times by the string? (a) Plane curve; (b) closed curve, not simple; (c) simple closed curve; (d) curve, not closed; (e) closed curve; (f) a triangle.

11. a) A bicycle tire and a coin help children to distinguish between two geometric concepts. What are they?
    b) A narrow rectangular picture frame and a sheet of typing paper help children distinguish between two concepts. What are they?

12. Name objects that you might find in a classroom that show plane figures, such as a line segment, a right angle, a closed curve, a square, and a triangle.

13. Name objects that you might find in a classroom that show space figures, such as prisms, cubes, spheres, pyramids, and so on.

1–5
movements in space

We handle, observe, and study a variety of physical objects, including models of figures, and then describe and classify the figures suggested by the objects. Then by moving the models in various ways we can discover many properties of the figures that the models show.

In this chapter we take an intuitive view of "movements of figures" as we move models of figures. Later, in Chapter 4, we shall explain the concept of motion naturally in terms of functions.

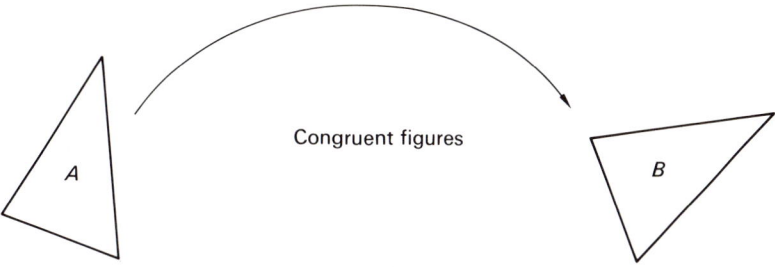

Congruent figures

The basic assumption we make when figures are moved in space (an assumption that is intuitive to children) is that the figure remains rigid as it is moved. Thus, if the figure above is moved from position A to position B without any deformation, the figure in position A is said to be congruent to the figure in position B.

parallel lines and parallel movement The notion of parallelism of lines is an intuitive concept that children, after some experience

20 informal geometry

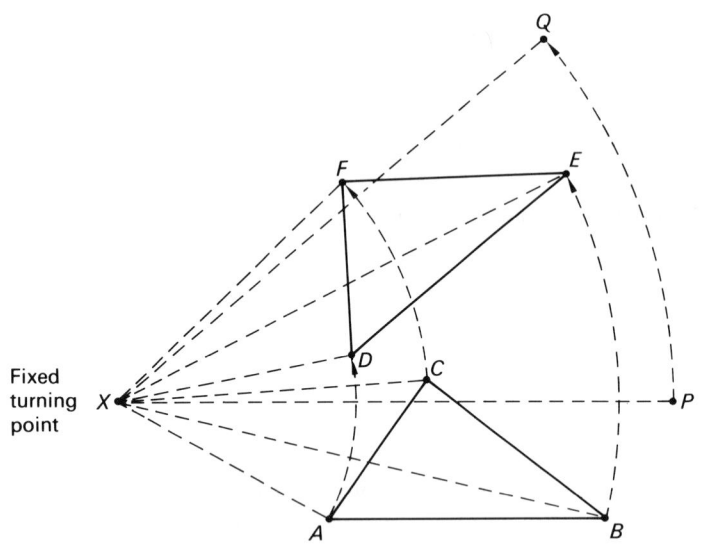

Fig. 1–15. △ABC is moved to △DEF by a turning movement about point X.

C is moved to F. △ABC is congruent to △DEF. This same turning movement takes each point P of the plane to some point Q, so that every point (except X) is moved along a circle with center at X.

Experiments with paper models of plane figures show that parallel and turning movements can be applied to a figure, one after another, any number of times, and that the figure in the new position will be congruent to the original one.

folding movements We might be tempted to decide that, given two congruent figures in a plane, one of the figures can be moved to the other by a series of parallel and turning movements. But consider the congruent triangles ABC and DEF in Fig. 1–16, and try to move a copy of one triangle to the other. No matter how we turn or slide one figure in the plane, it will not fit the other.

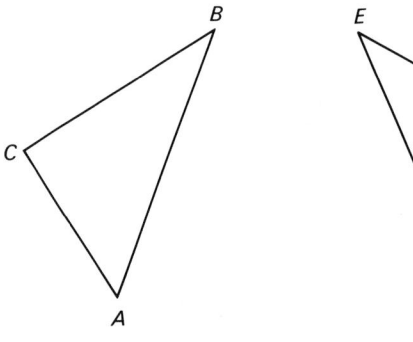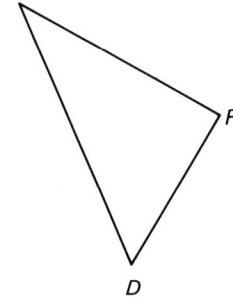

Figure 1–16

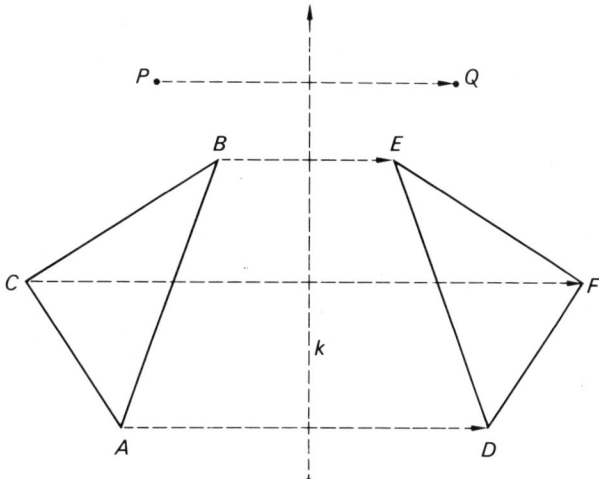

Fig. 1–17. △ABC is moved to △DEF by folding movement about line k.

Yet there is a motion that will move one of these triangles to fit the other. We can trace the two triangles on thin paper, and when we fold the paper along line k we note that △ABC will fall on △DEF (Fig. 1–17). Of course, this motion is not restricted to the plane of the triangles, since the folded paper was lifted off the plane. The folding movement can be thought of as a plane motion by imagining a mirror along line k. Then △DEF is the mirror image of △ABC, or the reflection of △ABC in line k. The folding movement about line k has the effect of moving each point P of the plane to some point Q along a line perpendicular to k. (The points of k are "moved" into themselves.)

Thus we agree that any combination of parallel, turning, and folding movements of a plane figure yields a congruent figure, and conversely, any figure can be moved to any other congruent figure by a series of these movements.

turning and folding symmetries Plane movements that preserve the size and shape of a figure—that is, that preserve congruence—are called *rigid movements*, or *isometries*. We can apply these movements to parts of given figures as a means of further classifications of the figures. We do so by means of the notion of symmetry. If a rigid movement of a figure takes the figure into itself, we say that the movement is a *symmetry* of the figure. For example, if a line segment $\overline{AB}$ is turned about its midpoint one-half turn, the segment is moved into itself with B on A and A on B. Thus $\overline{AB}$ has a turning symmetry about its midpoint. Also, a folding movement about the diagonal of a square moves the square into itself. Thus a square has a folding symmetry about a diagonal.

As we examine and move figures, we discover various consequences of symmetries and congruences. Some of the discoveries are:

> There is no parallel symmetry possible for any figure except a straight line.
>
> Triangles (Fig. 1–18) can be classified according to their folding symmetries: A triangle has 0, 1, or 3 folding symmetries.

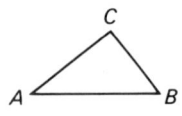 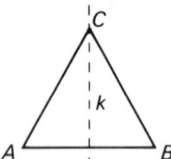 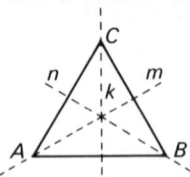

Triangle.
No folding symmetries

Isosceles triangle.
Exactly one folding symmetry, along k; $\overline{AC}$ is congruent to $\overline{BC}$

Equilateral triangle.
Exactly three folding symmetries, along k, m, n; all three sides congruent

Figure 1–18

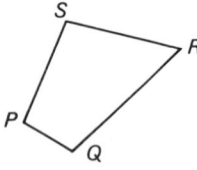 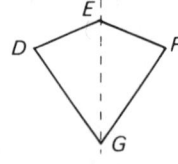

Quadrilateral.
No folding symmetries

Quadrilateral.
Exactly one folding symmetry; two pairs of adjacent, congruent sides

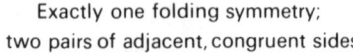

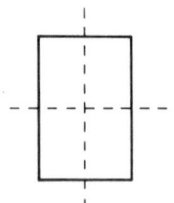 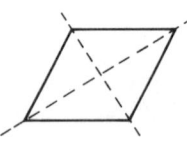 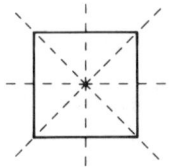

Rectangle.
Exactly two folding symmetries; two pairs of opposite, congruent sides

Rhombus.

Square.
Exactly four folding symmetries; all sides congruent

Figure 1–19

> Quadrilaterals (Fig. 1–19) can be classified according to their folding symmetries: A quadrilateral has 0, 1, 2, or 4 folding symmetries.

An equilateral triangle (Fig. 1–20) has three turning symmetries about the point of intersection of its folding lines.

A rectangle (Fig. 1–21) has two turning symmetries about the point of intersection of its folding lines.

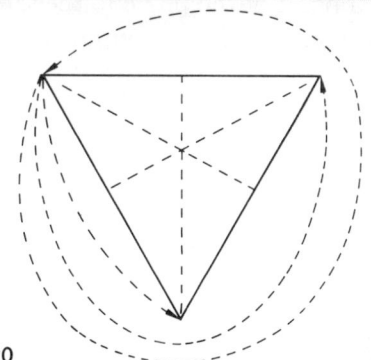
Figure 1–20

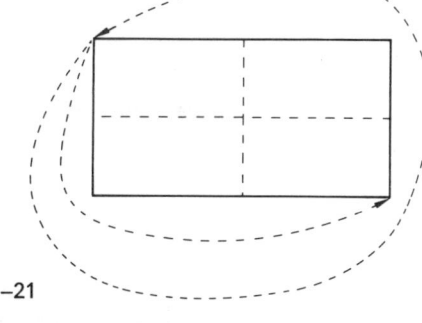
Figure 1–21

A square (Fig. 1–22) has four turning symmetries about the point of intersection of its folding lines.

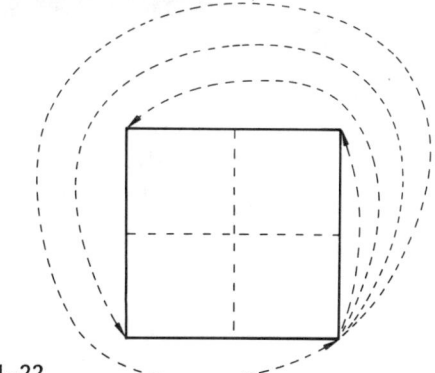
Figure 1–22

A circle has any number of folding and turning symmetries. (Children can use this idea to help them distinguish between circles and other oval figures that appear almost circular.)

Remark When symmetries of a figure are counted, we of course count the number of *different* symmetries of the figure. Two symmetries of a figure are different if and only if there is at least

24   informal geometry

one point of the plane whose image under one of the symmetries is different from its image under the other symmetry. Thus, for example, the symmetry of a square that rotates the plane 180° clockwise about its center is not different from the symmetry that rotates the plane 180° counterclockwise about its center. Both of these symmetries take each point of the plane to the same image point, and thus they are considered the same symmetry.

exercise set 1–2

1. Draw a line $\overleftrightarrow{MB}$ and point P not on the line. Construct a line through P parallel to $\overleftrightarrow{MB}$ by use of parallel movements.

2. a) Draw a triangle ABC. Use parallel movements to construct a line through B parallel to $\overline{AC}$, a line through C parallel to $\overline{AB}$, and a line through A parallel to $\overline{BC}$.
   b) Name the points of intersections of the lines you constructed in (a) F, G, and H. What kind of figure is FGH?

3. Draw a rectangle $ABCD$. Turn the rectangle around point A so that B goes to M, C goes to T and D goes to S and no side of $ABCD$ intersects $AMTS$ except at A.

4. Describe a series of movements that takes triangle ABC to triangle MNP so that A goes to M, B to N, and C to P (Fig. 1–23). (Use a tissue paper model of ABC.)

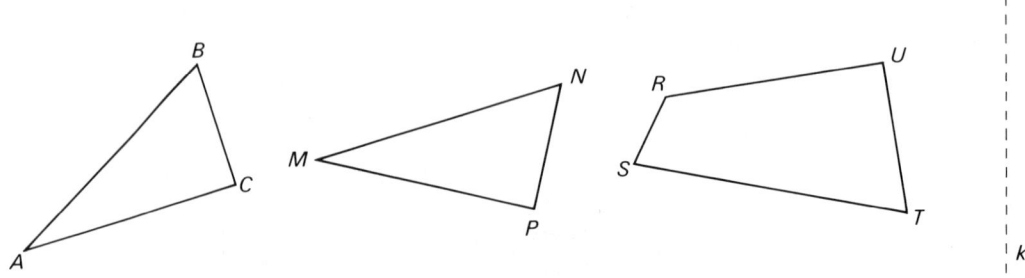

Figure 1–23   Figure 1–24

5. Draw $RSTU$ and line k, as shown in Fig. 1–24. Move $RSTU$ by a folding movement about k to a figure $ABCD$.

6. Describe a series of movements that takes triangle ABC to triangle DFE so that A goes to D, B to F, and C to E (Fig. 1–25).

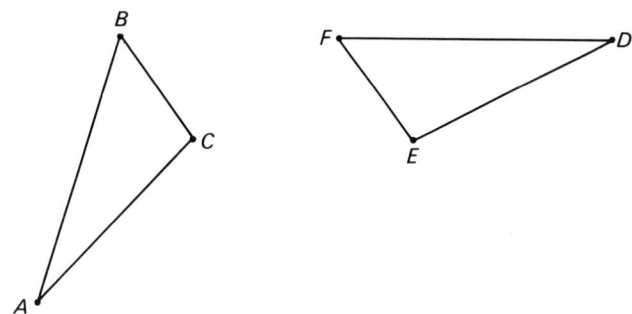

Figure 1–25

7. a) How many folding symmetries has a square?
   b) How many turning symmetries?

8. a) How many folding symmetries has a rectangle?
   b) If a quadrilateral has two folding symmetries, can we deduce that the figure is a rectangle?

9. Is it possible for a quadrilateral to have exactly three folding symmetries?

10. (a) Is it possible for a pentagon to have exactly one folding symmetry? (b) Exactly two folding symmetries? (c) Exactly three? (d) Exactly four? (e) Exactly five?

11. a) How many folding symmetries has a circle?
    b) How many turning symmetries?

12. Is it true that: If any polygon of n sides has r folding symmetries, then the number r is a factor of n?

13. Use parallel, turning, or folding movements to test whether the two pentagons in Fig. 1–26 are congruent. Describe your method.

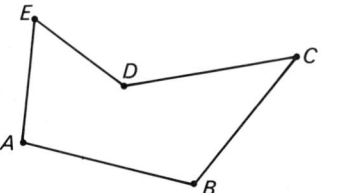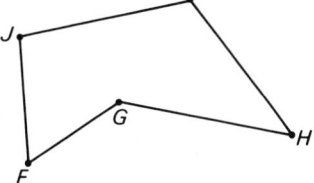

Figure 1–26

14. What movement takes one figure onto the other in each of the figures in Fig. 1–27?

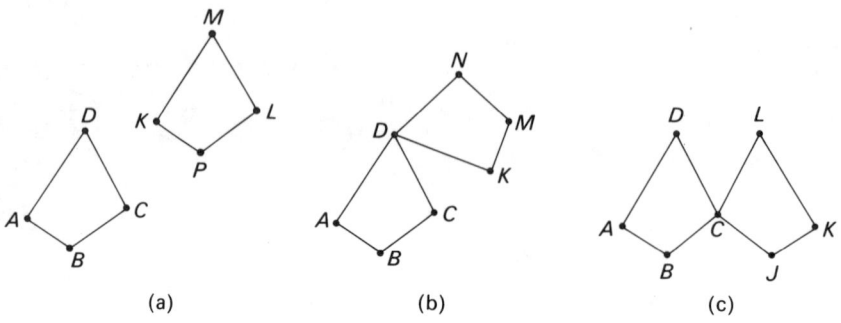

(a)   (b)   (c)   Figure 1–27

15. a) What movements take flag l to flag q in Fig. 1–28?
    b) Describe two sequences of movements that take flag q to flag l. Draw figures as needed.

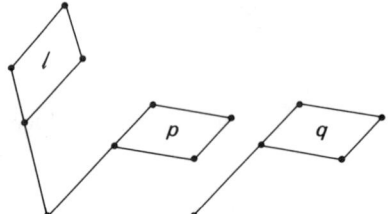

Figure 1–28

16. In Fig. 1–29, $\triangle ABC$ was moved by a parallel movement to $\triangle MPQ$ and then $\triangle MPQ$ was moved by a parallel movement to $\triangle RST$. What single movement takes $\triangle ABC$ to $\triangle RST$?

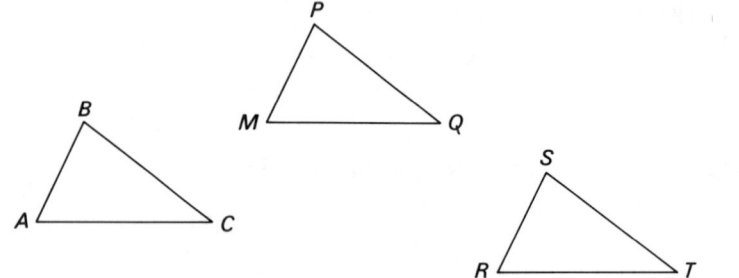

Figure 1–29

17. In Fig. 1–30, $\triangle ABC$ was moved by a folding movement to $\triangle EDF$; and $\triangle EDF$ was moved by a turning movement about F to $\triangle HGF$. Can $\triangle HGF$ be moved to $\triangle ABC$ by one movement? If not, describe how $\triangle HGF$ can be moved to $\triangle ABC$ by a parallel movement followed by a folding movement.

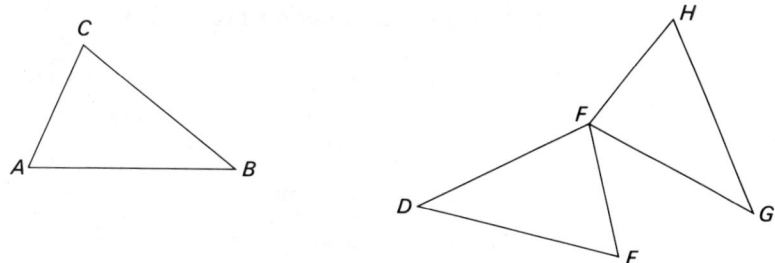

Figure 1–30

18. a) How many folding symmetries has each of the figures in Fig. 1–31?
    b) How many turning symmetries?

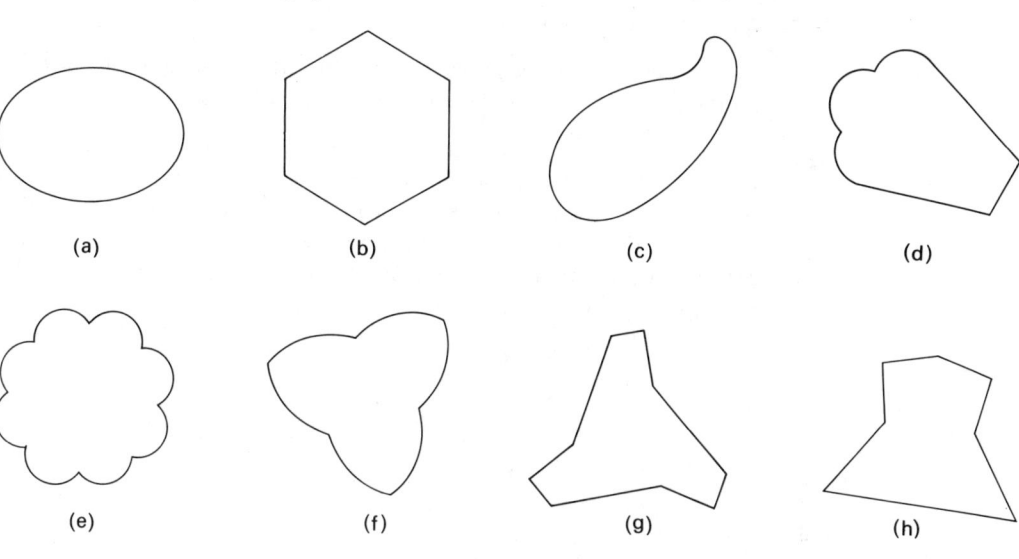

Figure 1–31

19. Name the quadrilateral which fits each of these descriptions.
    a) Only 2 folding symmetries are about its diagonals.
    b) Only 2 folding symmetries are about lines that bisect and are perpendicular to pairs of opposite sides.
    c) Four folding symmetries are about its diagonals and the perpendicular bisectors of pairs of opposite sides.

20. Does the parallelogram below have folding symmetries?

Figure 1–32

teaching questions and projects 1–2

1. List pairs of letters of the alphabet (capital and lower-case) that can be transformed into one another by a series of movements. Describe the movements.

2. Do the same for numerals.

3. Devise a game or puzzle for children that uses mirror images or paper folding.

4. Plan an activity using paper cutouts for classifying quadrilaterals according to lines of symmetry.

5. Plan an activity using paper cutouts to find the symmetries of (a) a rectangle, (b) an equilateral triangle, and (c) a square.

6. Suppose that □ is a unit square. You can form larger figures by putting two units together so that an edge matches an edge. How many *different* plane figures can you form using 4 units? 5 units? ("Different" means that they cannot be made to coincide under a series of rigid motions.) Plan an activity using these and similar questions to develop the idea of congruence.

chapter two

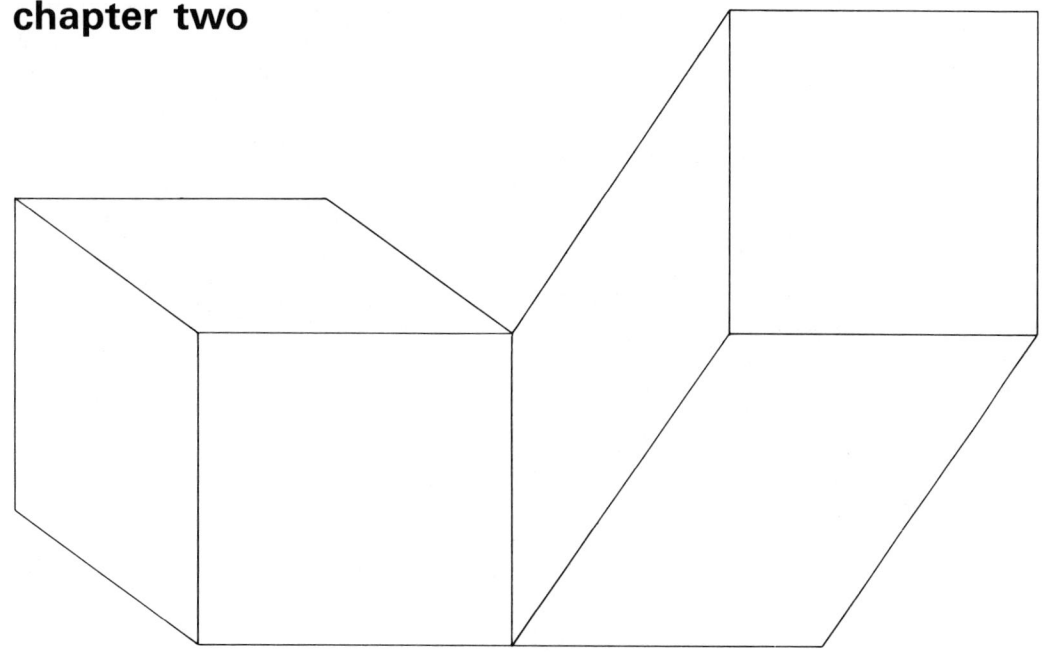

measurement of geometric figures

2-1
introduction

As a result of experiences, all of us learn to classify objects around us not only according to shape but also according to size. The progressive stages through which we learn to recognize and utilize certain properties of shapes of geometric figures—such as straightness, flatness, and roundness—were discussed in Chapter 1. Knowing these properties leads us to classify figures and identify various relations among the properties themselves. We can also classify figures according to size.

30 measurement of geometric figures

Sizes of figures are often compared by moving copies of the figures around on a plane or in space. An intuitive concept of congruence results from an awareness that one figure can be made to fit exactly on another; that is, two congruent figures have the same shape and size.

For example, you might settle the question about which of two pencils is the larger by holding the pencils upright and side by side, with the ends on a table top. You assume that the pencils have the same shape; you are concerned only about size. As a result of comparison you conclude, "One pencil is longer than the other" or "One pencil is as long as the other."

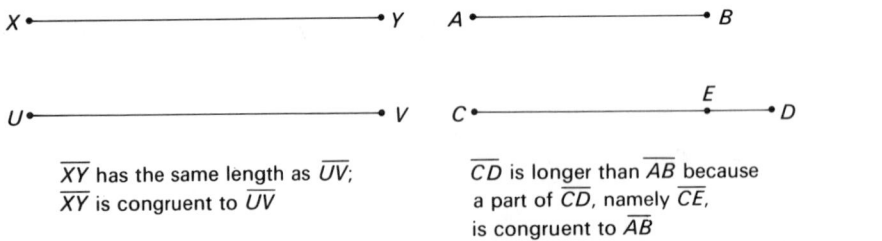

$\overline{XY}$ has the same length as $\overline{UV}$; $\overline{XY}$ is congruent to $\overline{UV}$

$\overline{CD}$ is longer than $\overline{AB}$ because a part of $\overline{CD}$, namely $\overline{CE}$, is congruent to $\overline{AB}$

Figure 2–1

At this point, our intuitive concept of length is only a comparative one (see Fig. 2–1). Two line segments either: have the same length (they are congruent); or one is longer than the other (a part of one is congruent to the other).

Our first intuitive concept of length is a correct one: All lengths are comparisons with a given segment. Thus the word length cannot be defined in isolation. Only when a particular line segment is specified as a unit of comparison can we say that another line segment is "as long as the unit segment" (congruent to the unit segment) or "longer than the unit segment." Then the length of a segment $\overline{PQ}$ is a measurement which indicates a unit segment and the number of such units needed to "cover" $\overline{PQ}$.

We make decisions about the sizes of other geometric figures in the same way. You could compare the lengths of your pencils because you agreed tacitly that the pencils have the same sort of shape. We would hesitate to compare the size of a pencil and a sheet of paper, for example, because these objects have different kinds of shapes. But we could consider the problem of deciding which of two sheets of paper is the larger. We think of each of these sheets as a rectangular region and compare the sizes of the regions.

Let us assume that we are comparing two regions A and B shown in Fig. 2–2(a). As children, we would make a comparison by placing B on top of A (Fig. 2–2b). But then each of us could argue that a part of the other's figure is "left over." After some experimentation, we might solve the problem by cutting B into two pieces and rearranging the pieces so that they fit inside A (Fig. 2–2c). In this example we find that A is larger than B because a part of A is left uncovered by this rearrangement of B. Note that the process of fitting one plane region inside another region can be accomplished by separating one region into smaller regions and assembling them so that the pieces do not overlap.

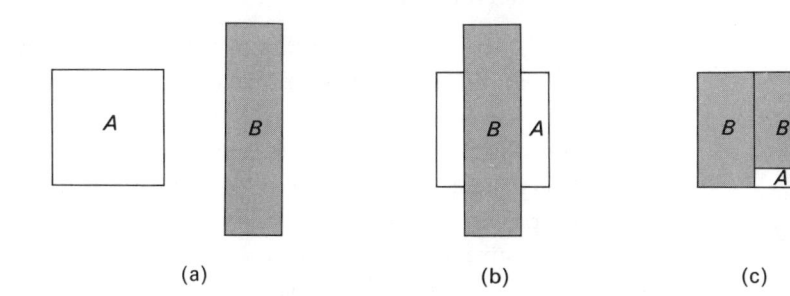

Figure 2–2

(a) (b) (c)

At a slightly higher level of maturity, the same process can be repeated in comparing the sizes of two space figures. Consider, for example, the comparison of a space figure suggested by a shoebox and another space figure in the form of a cube of styrofoam. By cutting the cube of styrofoam into pieces and packing the pieces in the shoebox, we can find whether or not the box is completely "filled" by the cube and decide whether the box is larger or smaller than the cube of styrofoam or whether they are the same size.

Such experimental activities in comparing sizes of figures lead to some informal agreements that we refine and state as basic principles for comparison of figures:

Only geometric figures of the same kind can be compared for size. (Segments can be compared with segments, plane regions with plane regions, and space figures with space figures.)

Two figures of the same kind have the same size if the figures are congruent.

One figure is larger than a second figure of the same kind if a part of the first figure is congruent to the whole second figure (or to some nonoverlapping rearrangement of all the parts into which the second figure is separated). (See Fig. 2–3.)

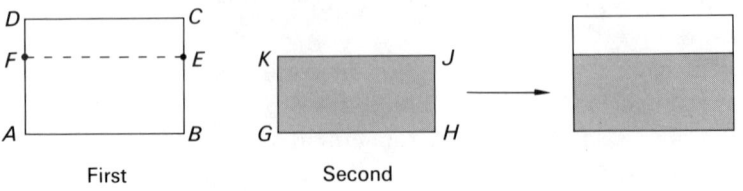

Fig. 2–3. First region is larger than second because a part of first, ABEF, is congruent to whole second region, GHJK.

2–2
the measure of a figure

When we compare lengths of pencils and sizes of sheets of paper, we usually are free to move the pencils and papers and cut the papers into pieces. This allows the possibility of making direct comparisons of lengths of segments and of sizes of regions. A higher level of thinking about comparison results when we are confronted with a situation in which the objects cannot be moved. For example, suppose we need to know which is longer: a curtain rod on one side of a room or a towel rod on the other side of the room. Neither can be moved. We decide to compare each rod with some object that *can* be moved. Suppose an available pencil is chosen as the movable figure. We find out how many of these pencils are in the length of each rod; we place the pencil along each rod, with each new position of the pencil starting at the end of the former position (possibly marking the pencil lengths on the rods as we go). We report: (1) The towel rod A is a little longer than 3 pencils. (2) The curtain rod B is a little longer than 4 pencils. (3) Since 4 is greater than 3, the curtain rod B is longer than the towel rod A; in fact, rod B is longer than rod A by about 1 pencil. (See Fig. 2–4.)

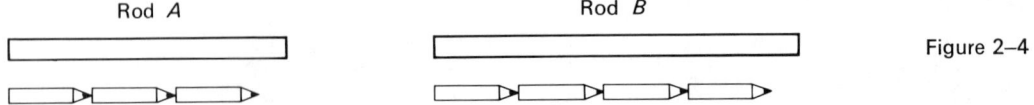

Figure 2–4

In situations similar to this we made our first discoveries about the measurement of a line segment. Later we used more precise language, but the principle remains the same. In the above situation a pencil is the *unit of measure;* it is the segment used as a standard of comparison. The number 3 is the *measure* of segment A in terms of this unit of measure. These two pieces of information together describe the length of the segment:

$$\underset{\underset{\text{(Measure)}}{\uparrow}}{\text{Length of rod } A = 3} \quad \underset{\underset{\text{(Unit of measure)}}{\uparrow}}{\text{pencils}}$$

Imagine that two children measure rod A with their own pencils and find different measures of the rod. For example,

$$\underset{\underset{\text{(Measure)}}{\uparrow} \downarrow}{\text{Length of rod } A = 6} \quad \text{of} \quad \underset{\underset{\text{(Unit of measure)}}{\uparrow} \downarrow}{\text{Mary's pencils}}$$

$$\text{Length of rod } A = 3 \quad \text{of} \quad \text{John's pencils}$$

The children at first may claim that rod A is twice as long when Mary measures it as when John measures it. But they will decide after discussion that the length of the rod does not change; the measuring unit is changed. Apparently Mary's pencil is shorter than John's. Comparison shows that

Length of John's pencil = 2 of Mary's pencils.*

To measure the segment suggested by the edge of rod A we note (or mark) the pencil lengths on the segment, as shown in Fig. 2–5. The points X, Y, Z on the line segment $\overline{UV}$, showing rod A, form three segments $\overline{UX}$, $\overline{XY}$, and $\overline{YZ}$, all congruent to $\overline{UX}$ (or to our pencil). The number of congruent segments, end to end, that are needed to "cover" the segment approximately, that is, to the nearest unit segment, is the measure of $\overline{UV}$ with respect to the unit of measure $\overline{UX}$. Thus it is meaningless to report, "The length of rod A is 3." We need to include the unit of measure as well as the measure and report, "The length of $\overline{UV}$ is 3 times the length of $\overline{UX}$."

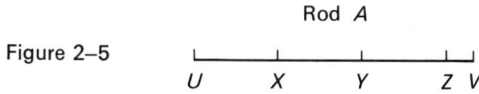

Figure 2–5

---

* The symbol "=" is used in this and similar sentences to mean that the two reported measurements are for the same figure, in this case the same segment.

34 measurement of geometric figures

Remark: The physical process of measurement is always approximate, for several reasons. First, the act of laying off a copy of the unit is always subject to human error. This is not a mathematical concern. Second, no matter how small a unit we choose when we measure an object, the number of units required seldom comes out even. We are then content to round off to the nearest unit and report this number of units as the measure of the object. In the case of measuring a segment such as $\overline{UV}$ with respect to unit $\overline{UX}$, the question of the "nearest unit" is answered by comparing $\overline{ZV}$ with $\overline{VW}$ in Fig. 2–6. Since $\overline{ZV}$ is shorter than $\overline{VW}$, we report the $\overline{UX}$-measure of $\overline{UV}$ as 3. If $\overline{ZV}$ and $\overline{VW}$ were congruent, or if $\overline{ZV}$ were longer than $\overline{VW}$, it would be the usual convention to round off to the greater measure and report the $\overline{UX}$-measure of $\overline{UV}$ as 4.

|←Unit→|←Unit→|←Unit→|←Unit→|

U   X   Y   Z V   W   Figure 2–6

A comparable situation for the measurement of plane regions is: Which is larger, a window glass or the door glass in a room? The objects cannot be moved. Children confronted with this problem might compare each glass with a plane surface that can be moved, such as a square piece of cardboard. They make many cardboard cutouts that are congruent to the small square that they choose as a unit. Then they lay the square cutouts edge to edge against the window glass and the door glass so that the squares approximately cover the regions without overlapping, as in Fig. 2–7. We can follow a similar plan mentally without applying the cutouts. Counting the squares, we find that region A is about 12 unit squares and region B is about 6 unit squares. Thus region A is larger than region B by about 6 unit squares.

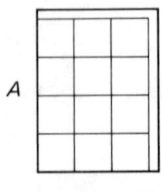

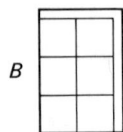

Unit square cutout   Window glass covered with square cutouts   Door glass covered with square cutouts   Figure 2–7

The measurement of a region is called its *area*. Again two pieces of information are needed to describe the area of a region. For example:

The area of region A = 12   unit squares
                          ↑   ↑
                     (Measure)   (Unit of measure)

Each region is compared with a unit region (usually a unit square region), and the area is reported by stating the unit of measure and the number of units that cover the region, to the nearest unit.

In the case of measuring plane regions, deciding the nearest unit is not as direct a process as it is in the case of line segments. With line segments, we can usually tell at a glance which endpoint of a unit segment is closer to the endpoint of the segment being measured, but it is not so simple with plane regions. For example, in Fig. 2–8 we might too hastily say that the area is 12 square units, to the nearest square unit, because the sides of the region are 3 and 4 units long, to the nearest unit segment. But if we collect the pieces of the region left over after we cover part of the figure with 12 unit squares, we see that these pieces form almost 2 of the unit squares. Thus the area is actually 14 square units, to the nearest square unit.

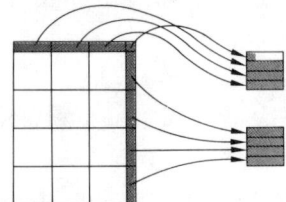

Figure 2–8

The same principles apply to the comparison of sizes of solid objects. How can children decide which of two open immovable boxes of different sizes is larger? Since they cannot fit one box inside the other, they can select a small space figure that can be moved and then compare each of the boxes with this unit space figure. They will find that a small cube is the most convenient unit because cubes fit against each other neatly. For example, they may select cubes of sugar and fill each of the two boxes with these unit cubes. They then count the number of cubes needed to fill each box and from this information decide which of the two boxes is larger.

36 measurement of geometric figures

The volume of box A = 24 cubes
             ↑   ↑
     (Measure)(Unit of measure)

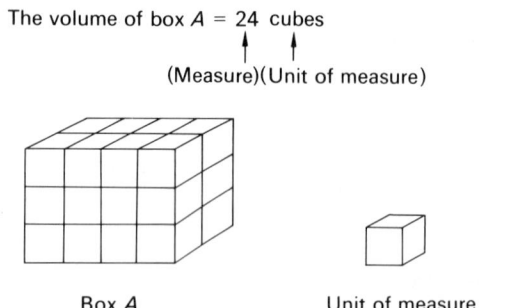

Box A   Unit of measure   Figure 2–9

The measurement of a space figure is called its *volume*. Again two pieces of information are needed to describe the volume of a space figure (see Fig. 2–9). Each space figure is compared with a unit figure (usually a unit cube), and the volume is reported by stating the unit of measure and the number of units that fill the figure, to the nearest unit. From experiences like these some *principles of measurement* can be identified:

> To measure a figure (a segment, a region, an angle, a space figure), choose as a unit of measure a figure of the same kind (a unit segment, a unit region, a unit angle, a unit space figure). Put together congruent copies of the unit (end to end, edge to edge, side to side) until they cover or fill the figure, to the nearest unit. The number of copies of the unit used is called the *measure* of the figure, in terms of the unit of measure.
>
> The unit of measure remains fixed in size as it is moved or as time passes.
>
> If two figures are congruent, they have the same measure when measured with the same unit.
>
> If a figure is separated into nonoverlapping parts, then its measure is the sum of the measures of its parts, all measured with the same unit.

2–3
standard units of measure

The history of the evolution of modern standard units of measure is an interesting one. Each village and tribe was forced by necessity to adopt standard units locally; then as trade flourished between localities the conflicting units needed to be resolved. Finally, in a world closely knit in commerce and science, there are standard units accepted world-wide, though it is still not unusual to find some unit of measure that is not standard.

As they learn to measure, children in a sense repeat this evolution. During their first activities of learning to measure segments, regions, and space figures, some uncertainty results. Various children reported the same distance—say across a room—to have different measures. A discussion of these differences may be used to advantage to help the children trace the cause to the differences in the lengths of the units of measure used. As the children think about the situation in their own way, they decide that they can avoid the difficulty if everyone chooses the same unit—a standard unit. In the discussions about possible standard units, children can make conjectures about the origin of the foot as a standard unit of length, about the relative sizes of the standard units of length, and about the desirability of a square unit of area rather than a round or rectangular unit.

At various stages in our education, we are introduced to devices in which copies of standard units of measure are put together, such as a foot ruler, a measuring tape, and a protractor. We shall discuss instruments for measuring later.

Some common standard units of measure that are used and compared are:

Units of length	Units of area	Units of volume	Angle unit
inch	square inch	cubic inch	degree
foot	square foot	cubic foot	
yard	square yard	cubic yard	
mile	square mile	cubic centimeter	
centimeter	square centimeter	cubic meter	
meter	square meter		

2–4
measuring line segments and lengths of circles

Any line segment can be measured in terms of a given unit of length, say an inch. The measure of the segment is then a number which tells how many inches are in the segment, to the nearest inch. Three facts emerge:

Given a unit of length, for each line segment there is exactly one number which is the measure of the segment with respect to the unit.

38   measurement of geometric figures

> No matter how small a unit of length we choose, the measure (number) of a segment will usually be approximate, in the sense that this number of units will not fit the segment but will be the closest number of units to the length of the segment.
>
> A statement such as
>
> $$24 \text{ inches} = 2 \text{ feet}$$
>
> means that the two measurements both describe the length of the same line segment.

We find lengths not only of line segments; we may also find lengths of some curves that are not line segments, such as circles. To measure lengths of curves, we must imagine a unit of length that is an arc that fits the curve. A flexible tape measure provides a model of such curved units of length.

As we measure lengths of circles (*circumferences* of circles) and lengths of diameters of circles, we note a relation between these two measures. Children may experiment to find what this connection is. For example, each pupil can choose a circle of any diameter, and with any unit of length find the measure d of the diameter and the measure c of the circumference with respect to this unit. Children suggest various ways to measure the circumference; some children may try to use a flexible tape measure; some may lay a piece of string along the circle and then measure the string; others may make a cardboard copy of the circle and roll the disc along a line until a marked point on the disc makes one revolution, and then measure the distance rolled (see Fig. 2–10).

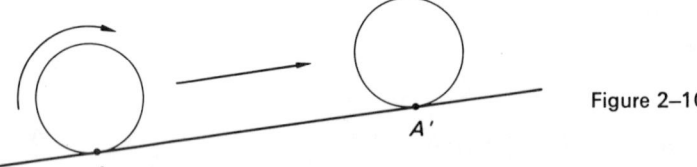

Figure 2–10

In all cases, the quotient $\dfrac{c}{d}$ will be slightly more than 3, and will appear to be constant regardless of the sizes of the circles and the units of length used. In an elementary school program, the best that can be done is to explain that this number is constant, is not a rational number, and is denoted by π (read "pi"), with value 3.1416, to the nearest .0001. Sometimes we approximate π

by $\frac{22}{7}$, which is in error by about .001. (We shall discuss irrational numbers in Section 2–9.) Experiences like this lead to the general formula for the circumference of any circle:

$$c = \pi d,$$

where c is the number of units in the length of the circle and d is the number of the same units in its diameter.

Remark: It is interesting to note here that the problem of proving that

$$\frac{c}{d} \text{ is a constant ratio for any circle}$$

is much deeper than is commonly realized. At each stage of school mathematics we tend to accept the formula $c = \pi d$ and we suggest that its proof will come in a higher-level course. By the time students are studying the calculus, the reasoning becomes circular; based on the assumption of the formula $c = \pi d$, one defines the trigonometric functions, and using derivatives and integrals of these functions, one gives a "proof" of the formula $c = \pi d$. Indeed, a genuine proof is quite sophisticated and involves finding limits of sequences of lengths of inscribed and circumscribed polynomials of a circle. For example, one such approach (by Wallis) yielded the result

$$\pi = 2\left(\frac{2}{1} \cdot \frac{2}{3} \cdot \frac{4}{3} \cdot \frac{4}{5} \cdot \frac{6}{5} \cdot \frac{6}{7} \cdot \frac{8}{7} \cdot \frac{8}{9} \cdots \frac{2k}{2k-1} \cdot \frac{2k}{2k+1} \cdots\right)$$

as the limit of an infinite product.

Here we have an example of a mathematical formula, $\pi = \frac{c}{d}$, that can be recognized as plausible through simple experiments, but that must be taken on faith throughout an elementary mathematics program.

2–5
measuring an angle

An angle, formed by two rays with a common endpoint, cannot be represented entirely with a drawing, but the use of arrows indicates that the sides (rays) extend without end. The problem of representing an angle is nontrivial; unless a student understands the model, he can formulate an incorrect concept of an angle. However, children's imaginations are vivid, and when they understand the use of the arrows to show rays, they are not disturbed by the fact that only a part of an angle is shown on a figure.

The principles of the measurement of line segments, regions, and space figures also apply to the measurement of an angle. The unit of measure must be an angle. Suppose we are given an angle, such as $\angle XYZ$ in Fig. 2–11, to measure. We may select any angle for a unit, such as $\angle UVW$. To measure $\angle XYZ$, we decide how many congruent copies of the unit angle, placed side to side, will form $\angle XYZ$. To find the measure, we could make a paper model of the unit angle and trace the model successively on $\angle XYZ$, as shown in Fig. 2–11. In this example, the measurement of $\angle XYZ$ is 4 unit angles, to the nearest unit.

Copies of unit angle

Unit angle

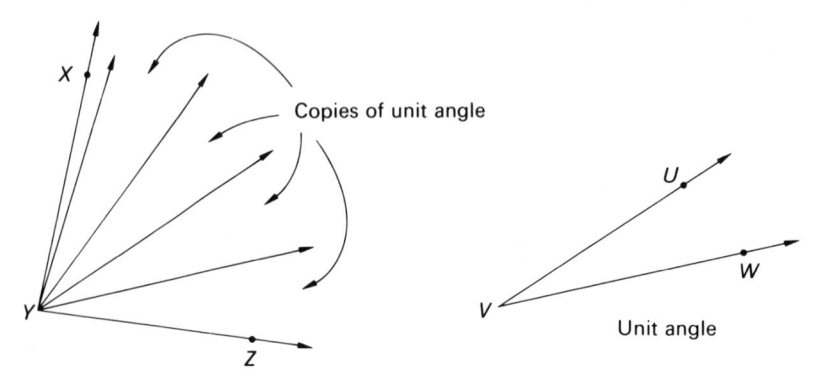

Figure 2–11

The measure of a given angle varies with different size units. Therefore, as in other forms of measurement, we need a standard unit. When children are aware of this need and seek a standard unit angle, they may decide at once that a right angle is too large to be convenient and suggest a part of a right angle. They may experiment with some part of a right angle and then find that a widely used unit is the *degree unit angle*, denoted by °.

A degree unit is an angle such that 90 degree units form a right angle. Thus a right angle has the following measurements in right angle units and in degree units:

1 right angle = 90 degree unit angles = 90°.

With increasing maturity children learn to use a protractor in measuring angles in degrees in the same way that they use a ruler to measure segments in inches.

2–6
measuring plane regions

Experience shows that a convenient unit for measuring plane regions is a square, because congruent squares fit together, any edge to any edge.

Fig. 2–12. An array of n rows of elements with m elements in each row has n × m elements in it.

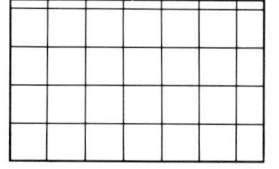

4 rows of squares with 7 squares in each row; area is 4 x 7 squares, to the nearest square.

The simplest plane region to measure is a rectangular region. We can build or imagine an array of unit squares that covers the rectangular region, to the nearest unit, and then find the required number of unit squares (see Fig. 2–12). To adults, the problem of finding the area of a rectangular region is an exercise in multiplication, a procedure that has been formalized. To children, the easiest way to find the area is to count the number of units covering the region. They soon have no need to do the actual placing of copies of the square unit on the rectangular region. Instead they measure two adjacent sides of the rectangle to find the number of rows of unit squares required and to find the number of unit squares used in each row. At this point they are ready to derive the formula for the area of a rectangular region even though the words and the symbols used may not be precise. They may say, for example, "The area of the rectangle is the length of the rectangle times its width." In more precise language this means "The measure of a rectangular region in square units (area of the rectangle) is the measure of a longer side in units (length) times the measure of a shorter side in units (width), where the same unit of length is used to measure each side of the rectangle, and the square unit has its sides each as long as this unit." Their first way of saying this becomes an abbreviation for the more precise statement. Thus in the formula

$$A = lw,$$

A represents the number of square units in the area, l the number of units in the length, and w the number of units in the width of the rectangle, all units being the same length as a side of the unit square.

In Fig. 2–12, a more careful measurement of the sides of the region is $4\frac{1}{4}$ units for the width and $6\frac{2}{3}$ units for the length. Then the area is $4\frac{1}{4} \times 6\frac{2}{3}$ square units, or $28\frac{1}{3}$ square units. In this way we verify our claim that the region has area 28 squares, to the nearest square.

An elementary stage in measuring any nonrectangular region is to cover the region with unit squares, to the nearest unit, and count the squares. The next stage in measuring is to place a grid on the figure, with the lines of the grid spaced a unit apart (see Fig. 2–13). We can then partly determine the number of squares by multiplication when we note rectangular arrays in the interiors of the regions; we do the rest of the counting unit by unit.

The third stage comes when we begin to speculate on ways to cut and rearrange plane regions to determine an area from other known areas. For example, $\triangle BCE$ in Fig. 2–13 can be shown to have exactly one-half the area of the rectangular grid $ABED$ which is placed on it. Children can discover this by cutting out triangles ABC and DEC and placing them on triangles FCB and FCE, respectively. Thus the area of $\triangle BCE$ is one-half the area of rectangle $ADEB$, that is, one-half the product of the length of the base $\overline{BE}$ and the height $\overline{CF}$. More examples of this sort give further evidence, and a general formula for the area of any triangle is suggested:

The area of a triangle equals one-half the product of the length of the base and the height of the triangle: $A = \frac{1}{2}bh$.

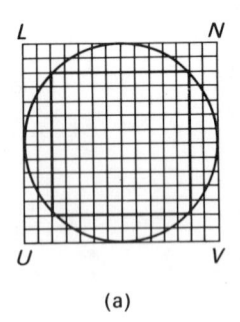

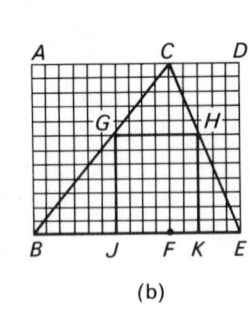

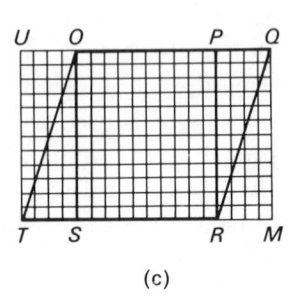

Fig. 2–13. (a) 10×10 squares in the array, about 54 more; area of *circle* is about 154 square units. (b) 7×6 squares in the array, about 48 squares more; area of *triangle* is about 90 square units. (c) 12×10 squares in the array, about 48 squares more; area of *parallelogram* is about 168 square units.

We shall prove this conjecture later when the meaning of proof is understood and when we have had some experience with congruent triangles (see Chapter 3). In a similar way a formula for the area of a parallelogram can be discovered by cutting triangle TOS out of the parallelogram $TOQR$ in Fig. 2–13 and rearranging the parts to form rectangle $OQMS$.

Similar experiences result in an approximation for the area of a circle. By using smaller and smaller unit squares and by refining the methods of counting the number of squares needed to cover the circle, we begin to suspect that a general formula could be found. A project like the following helps children in their attempt to find such a formula.

Project: Let each child use any unit of linear measure and choose a circle of any radius; find the measure r of the radius in this unit, the measure A of the circular region in square units, and the ratio $\dfrac{A}{r^2}$ of the area of the circle to the area of a square of side r. (The square unit must have its sides equal in length to the unit of linear measure.) The children will notice that all the ratios are about the same number, a number slightly greater than 3. This ratio appears to be constant for all circles, regardless of the unit of measure used; the number is the same number π that the children encountered in the formula for the circumference of a circle.

Experiences like this lead us to accept the general formula for the area of any circle:
$$A = \pi r^2,$$
where r is the number of units in the radius and A is the number of square units in the area.

2–7
measuring space figures

The method of finding the volume of a space figure is a direct extension of the preceding methods. When confronted with a problem of measuring a space figure, such as a box, we follow the principles of measurement and select a unit of measure that is also a space figure and proceed to "fill" the box with copies of this unit. We may decide after some experimentation that a cube is a convenient space unit, because congruent cubes fit

44   measurement of geometric figures

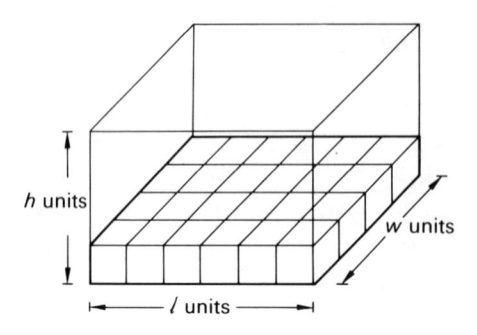

Fig. 2–14. Each layer contains $(\ell \cdot w)$ cubic units. There are h layers; thus the box contains $(\ell \cdot w) \cdot h$ cubic units.

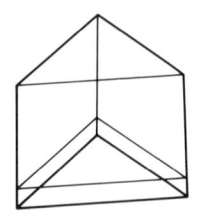

Fig. 2–15. Triangular prism.

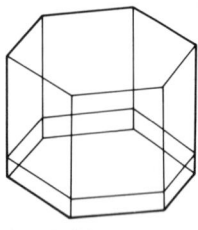

Fig. 2–16. Hexagonal prism.

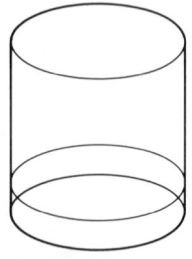

Fig. 2–17. Circular cylinder.

together neatly with no overlapping and no spaces between cubes left unfilled (Fig. 2–14).

The technique of counting the number of cubes that fill a box is also an extension of previous techniques. Each layer of cubes in the box is a rectangular array of cubes; we find the number of cubes in each layer by multiplying the number of rows by the number of cubes in each row; that is, the length l times the width w of the bottom of the box, in units of the same length as a side of the unit cube. The number of layers in the box is determined by the height h of the box. Thus there are h layers, with $l \cdot w$ cubes in each layer, giving a total of $(l \cdot w) \cdot h$ cubes in the box.

The formula for the volume of a box is

$$V = lwh,$$

where there are V cubic units in a box with length l units, width w units, and height h units, all units the same length as a side of the unit cube.

A *box* is a special example of a figure called a prism. It is a rectangular prism, because all its faces are rectangular regions. In general, a *prism* is any space figure with two congruent parallel faces in the shape of a polygon, and all other faces rectangular (see Figs. 2–15 and 2–16).

The same principles also apply to the measurement of the volume of any prism, in cubic units. The measurement results from thinking of a layer of cubes on the base of the prism and counting the number of layers needed to fill the prism.

The procedure for finding the volume of a prism applies also to a space figure with congruent bases that are circular regions. Such a figure is a *circular cylinder* (see Fig. 2–17). Using the formula for the area of a circle, we can find the number of cubic units needed to fill the bottom layer and count the number of such layers needed to fill the cylinder. Then the formula for the volume of a circular cylinder is $V = \pi r^2 h$.

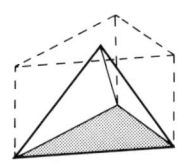

Fig. 2–18. Triangular pyramid. 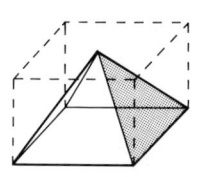   Fig. 2–19. Rectangular pyramid. 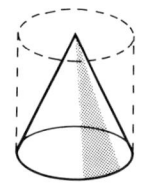   Fig. 2–20. Circular cone.

Space figures called pyramids (Figs. 2–18 and 2–19) are formed from prisms by connecting any fixed point on the top face of the prism to the points of the base of the prism. (The side faces of pyramids are triangular regions.) When a fixed point of the top face of a cylinder is connected to all the points of the base, the resulting figure is called a *cone* (Fig. 2–20). Experimentation of the type described for prisms and cylinders (or like those suggested in the exercises below) result in these generalizations:

The volume of a pyramid in cubic units is one-third the volume of its related prism.

The volume of a cone in cubic units is one-third the volume of its related cylinder.

exercise set 2–1

1. a) What is the length of $\overline{MN}$, using $\overline{AB}$ as a unit of measure?
   b) What is the length of $\overline{MN}$, using $\overline{CD}$ as a unit of measure?
   c) What is the length of $\overline{MN}$, using $\overline{EF}$ as a unit of measure?

   A•———•B

   C•——•D   M•————————————————————•N

   E•————•F

   d) What is the length of $\overline{MN}$, using an inch as a unit of measure?
   e) Explain the variation in the answers found in (a) and (d) with respect to the various units of measure.

2. a) What is the area of region Q, using region A as a unit of measure? (See page 46.)
   b) What is the area of region Q, using region B as a unit of measure?

46 measurement of geometric figures

c) What is the area of region Q, using region C as a unit of measure?
d) What is the area of region Q, using a square inch as a unit of measure?
e) Explain the variation in the answers found in (a) and (d) with respect to the various units of measure.

3. Determine units and measures to make each pair of measurements equal.

a) 1 inch	2.54 cm	g) 1 nautical mile	_____ feet
b) 1 mile	_____ furlongs	h) 1 nautical mile	_____ miles
c) 1 mile	_____ rods	i) 1 mile	_____ yards
d) 1 rod	$16\frac{1}{2}$ _____	j) 1 furlong	_____ yards
e) 1 yard	_____ cm	k) 1 section	_____ acres
f) 1 meter	_____ inches	l) 1 square inch	_____ square foot

4. A United States one-dollar note is six inches long; this is an approximate measurement. What is meant by "approximate measurement" in this sentence?

5. What marks are needed on a measuring stick if you use it to measure to the nearest (a) 1 inch? (b) $\frac{1}{2}$ inch? (c) 0.1 inch?

6. Describe the longest and the shortest segment that can have these measurements.
   a) 7 inches, to the nearest inch   b) 0.9 inch, to the nearest 0.1 inch
   c) $\frac{3}{8}$ inch, to the nearest $\frac{1}{8}$ inch

7. What is the largest possible error you can make if you measure a segment in these units of measure: (a) inch, (b) 0.1 inch, (c) $\frac{1}{8}$ inch?

8. Angles may also be measured in these units: radians, grads, and mils (π radians $= 180°$; 100 grads $= 90°$; 1600 mils $= 90°$). Complete these statements.
   a) π radians $=$ _____ grads $=$ _____ mils.
   b) 1 grad $=$ _____ radians $=$ _____ mils $=$ _____ degrees.
   c) 1 mil $=$ _____ radians $=$ _____ grads $=$ _____ degrees.

2–7   measuring space figures   47

Figure 2–21   (a)   (b)   (c)

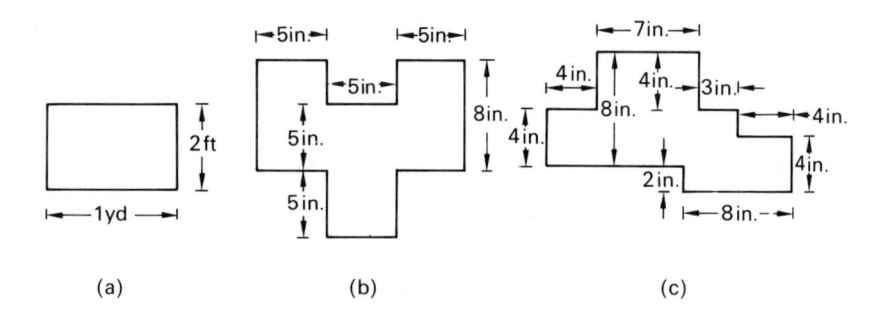

9. What are the areas of the regions shown in Fig. 2–21?

10. Given that the perimeter of a rectangle is 24, guess what its dimensions are for it to have an area as large as possible. Check your guess.
Try $l = 4$, $w = 8$. What is A?
Try $l = 5$, $w = 7$. What is A? Larger than before?
What measurements would you try next?

11. What happens to the area of a rectangular region if (a) l is doubled? (b) w is halved? (c) l is doubled and w is halved?

12. a) What happens to the area of a circular region if d is doubled?
    b) If d is halved?

13. Find the circumference and the area of the regions of circles with radii (a) 10.0 centimeters, rounded to the nearest 0.1 centimeter; (b) 3.700 meters, rounded to the nearest 0.01 meter; (c) 5.20 inches, rounded to the nearest 0.1 inch.

14. What are the volumes of the figures in Fig. 2–22?

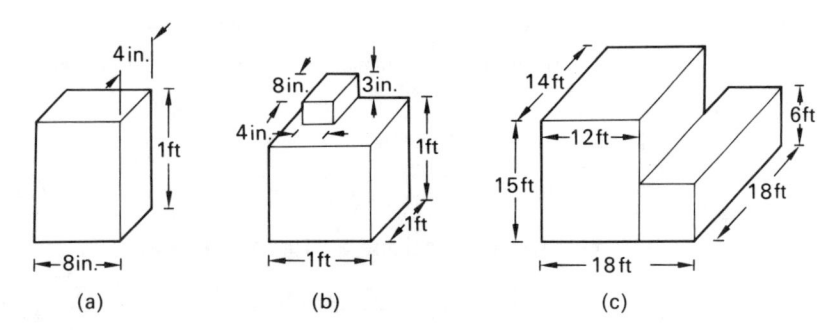

Figure 2–22   (a)   (b)   (c)

15. What happens to the volume of a rectangular prism if:
    a) l is doubled
    b) w is halved
    c) l and w are doubled
    d) l is doubled and w is halved
    e) l, w, and h are each halved
    f) l, w, and h are each doubled

48   measurement of geometric figures

16. Given that the twelve edges of a rectangular prism measure 48 in. altogether, guess what l, w, and h measure for the volume to be as large as possible. Check your guess.
Try $l = 3$, $w = 1$, $h = 8$. What is V?
Try $l = 2$, $w = 2$, $h = 8$. What is V?
Try $l = 3$, $w = 2$, $h = 7$. What is V? Larger than before?
What measurements would you try to make V even larger?

17. Using Wallis' formula, what is an approximate value of π obtained by letting $k = 8$?

18. Find the volume of each of the figures in Fig. 2–23.

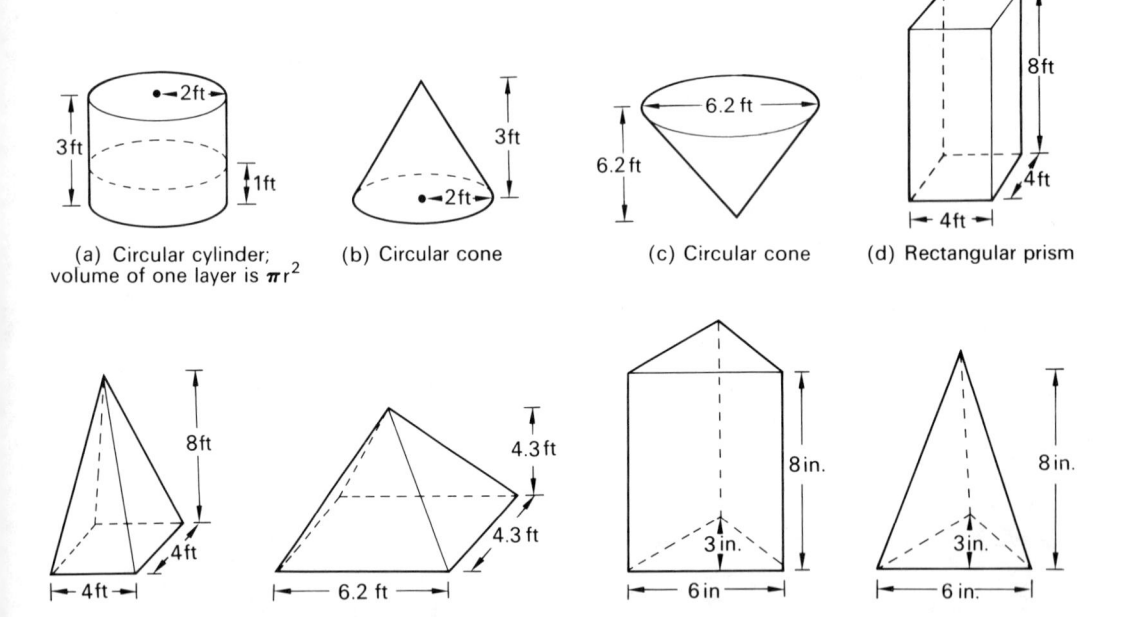

(a) Circular cylinder; volume of one layer is πr^2

(b) Circular cone

(c) Circular cone

(d) Rectangular prism

(e) Rectangular pyramid

(f) Rectangular pyramid

(g) Triangular prism

(h) Triangular pyramid

Figure 2–23

teaching questions and projects 2–1

1. a) How can young children decide which of these two segments is larger?

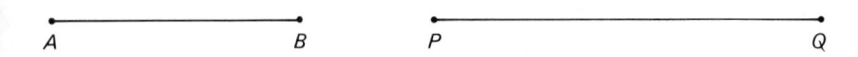

b) How can young children decide which of these two regions is larger?

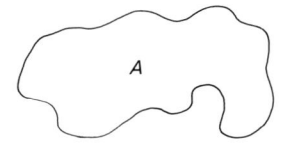

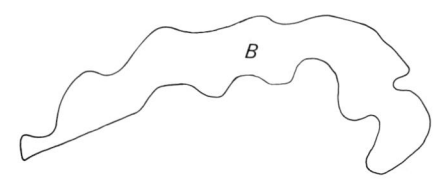

c) How can young children decide which of these two regions is larger?

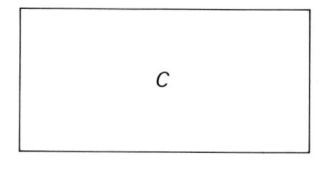

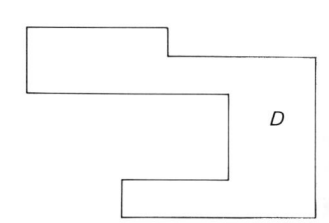

d) How can young children decide which of two tool boxes is larger?

2. a) Name some nonstandard units used to measure: length, area, and volume.
   b) What is the purpose of having children discuss or use nonstandard units of measure before standard units?

3. Name the standard units usually used to measure the following. (Consult a reference book if necessary.)
   a) The height of a person
   b) The height of a horse
   c) The length of a shoe
   d) The width of a shoe
   e) The size of a man's hat
   f) The distance to the sun
   g) The length of a football field
   h) The distance to a star
   i) The weight of a diamond

4. Consult an encyclopedia to find the origins of these units of length. What is the length of each in the foot unit of measure?
   a) inch
   b) foot
   c) yard
   d) rod
   e) fathom
   f) furlong
   g) mile
   h) nautical mile
   i) chain
   j) meter
   k) centimeter
   l) kilometer
   m) angstrom

5. Consult an encyclopedia to find the origin of these units of area. What is the area of each in the square mile unit of measure? (a) acre, (b) section, (c) township.

17. Develop a class activity in which children use objects such as chalkboard erasers to measure objects such as table tops, chalkboard edges. (For example, use long edge of eraser, then short edge of eraser and compare; use face of eraser to measure area.)

18. Describe how activities such as the above can be used to guide children to the discovery: $A = lw$.

19. Describe activities to help children discover formulas for the areas of parallelograms, trapezoids, and right triangles.

20. Prepare patterns that children may use to make models of space figures—prisms, pyramids, cylinders, and cones.

21. a) Describe an activity to help children discover the relation between the volume of a pyramid and its related prism and of a cone and its related cylinder.
    b) Extend the activities to help children discover formulas for the volume of a pyramid and a cone.

22. Describe an activity to help children discover formulas for finding volumes of space figures such as spheres, cones, cylinders. (For example, a chart such as this would help in finding the formula for the volume of a cylinder, $V = \pi r^2 h$.)

Cylinder	Radius, r	Height, h	Volume, V
A			
B			
C			

23. How can children measure volume by immersing objects in water?

24. Devise activities in constructing courts or fields for various games. (For example, a standard football field is 100 yards from goal line to goal line. It is marked off in units of 5 yards. Suppose that the field available for play is only 80 yards in length. Discuss ways to adjust or adapt the measurements to the field. What are the advantages and disadvantages of each way?)

25. Construct a treasure map in which different segments are measured in different units. Compare segments.

26. Devise puzzles in which paths measured with different units are compared.

27. Devise questions for children in which classroom measurements are compared to the measurements of other rooms (gymnasium, auditorium, cafeteria) and to the measurement of the entire building.

2–8
physical measurement and geometric measurement

A physical measuring process is seldom exact. No matter how finely the standard units are subdivided, the measure of a segment is still approximate, that is, to the nearest fractional part of the unit. If rulers are marked, for example, in sixteenths of an inch, the inch-measure of a segment is reported to the nearest sixteenth, and the difference between this measure and the "true" measure, whatever that may be, could be as much as $\frac{1}{32}$ inch. We say that this measurement has a possible error of $\frac{1}{32}$ inch, meaning that the round-off error is at most $\frac{1}{32}$ inch (see Fig. 2–27).

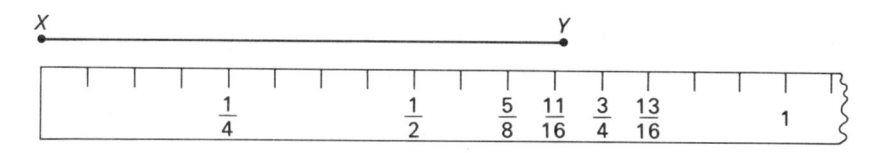

Fig. 2–27. Measure of segment $\overline{XY}$ in inches is $\frac{11}{16}$, with round-off error of at most $\frac{1}{32}$ inch. (Segment and ruler magnified.)

For practical purposes, in each physical measuring situation we agree on a certain amount of possible error and then subdivide the unit of measure into appropriately small parts. When we do this, we tacitly agree to ignore the resulting possible errors and report the measures of segments as rational numbers.

Remark: Let us review the definition and properties of rational numbers.

If a and b are integers, b not zero, then the number $\frac{a}{b}$ (or $a \div b$) is called a *rational number*.

$$\frac{a}{b} = \frac{c}{d} \quad \text{if and only if} \quad ad = bc.$$

Rational numbers are added and multiplied as follows:

$$\frac{a}{b} + \frac{c}{d} = \frac{ad + bc}{bd}; \quad \frac{a}{b} \cdot \frac{c}{d} = \frac{ac}{bd}.$$

Addition and multiplication of rational numbers are commutative and associative; multiplication is distributive over addition.

Zero is the identity element for addition; 1 is the identity element for multiplication.

For each rational number $\frac{a}{b}$ there is a rational number $\frac{-a}{b}$ such that $\frac{a}{b} + \left(\frac{-a}{b}\right) = 0$.

For each nonzero rational number $\frac{a}{b}$, there is a rational number $\frac{b}{a}$ such that $\frac{a}{b} \cdot \frac{b}{a} = 1$.

To return to what happens in a measuring situation: Assuming a given unit of measure that can be finitely subdivided, then for each line segment in the plane there is a positive rational number that is the physical measure of the segment with respect to the given unit.

The situation at this point in our development of measurement is analogous to the point in the description of line segments at which we thought of a segment as the physical edge of an object (such as a door) or as an object itself (such as a thin piece of wire) or as a drawing of the object. The next step in the description occurred when we abstracted the properties of the physical objects and considered an idealized or geometric line segment, a conceptual or mathematical model of the physical object. Then we took a drawing of the segment to be a representation of the idealized segment, but not the segment itself.

In the same way we extend our concept of measurement of segments from the physical measuring process to an idealized geometric measurement. In the physical sense, with respect to a given unit we assign to each segment a rational number, but because of round-off errors there may be segments of different lengths that are assigned the same rational number. In Fig. 2–28, $\overline{AB}$ is not congruent to $\overline{CD}$, but both have the same measure, 3, to the nearest unit. In the idealized sense of measurement, we assume that with respect to a given unit each segment has a measure, and if two segments are not congruent they have different measures.

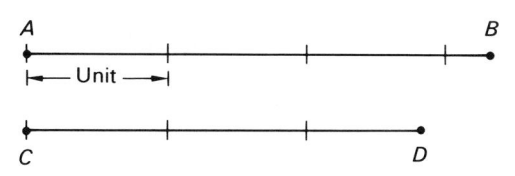

2–8   physical measurement and geometric measurement

Figure 2–28

To get an insight into geometric measurement, let us consider this example.

Example: Suppose we had a ruler marked in $\frac{1}{10}$ in., $\frac{1}{100}$ in., $\frac{1}{1000}$ in., $\frac{1}{10000}$ in., ... and imagine that with more and more powerful magnifying glasses we could read this ruler. Then imagine trying to measure a segment $\overline{AB}$ with this ruler. The first reading, for example, would be to the nearest $\frac{1}{10}$ in. Say point B is between $\frac{7}{10}$ and $\frac{8}{10}$ on the ruler. Then a magnifying glass is used to read the ruler to the nearest $\frac{1}{100}$ in. Say B is between $\frac{77}{100}$ and $\frac{78}{100}$ on the ruler. A stronger magnifying glass allows us to read the scale to the nearest $\frac{1}{1000}$ in. Say B is between $\frac{775}{1000}$ and $\frac{776}{1000}$ on the ruler. *Imagine that this process goes on without end* (see Fig. 2–29).

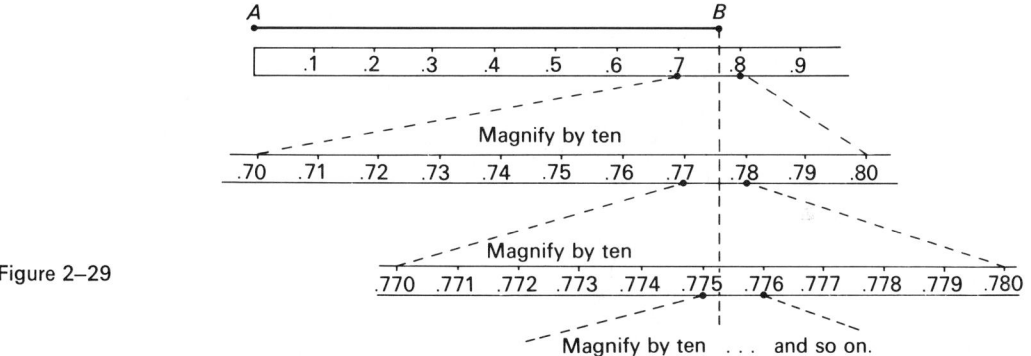

Figure 2–29

We can make conjectures about this process. There are two possibilities:

a) At some stage in the process, point B might fall on a division point of the ruler; then we call the measure of $\overline{AB}$ a rational number.

b) Point B might *never* fall on a division point at any stage in the process, no matter how far it is extended.

The first of these possibilities certainly can happen. Our assumption that the second possibility can also happen is a basis

56 measurement of geometric figures

for geometric measurement. As we shall see, this assumption leads to a set of numbers that (ideally) enables us to measure each segment and to assign different measures to any two segments that are not congruent. A formal statement of this assumption will be listed as a postulate in Chapter 3.

2–9
the real numbers

We agree that geometric measurement should assign different measures (numbers) to two segments that are not congruent. We also suspect that some of these numbers will not be rational numbers. Thus we need an appropriate set of numbers (which include the rational numbers as a subset) to serve as measures in geometric measurement.

Let us look back at the successive magnifications of the ruler that we used to measure segment $\overline{AB}$ in Fig. 2–29. We will say that the measure of $\overline{AB}$ is a number in our new set; call it n. Then n is between .7 and .8, between .77 and .78, etc. This can be written as:

$$.7 \leq n \leq .8$$
$$.77 \leq n \leq .78$$
$$.775 \leq n \leq .776$$
$$.7752 \leq n \leq .7753$$
$$\vdots$$

We say that n is in the interval [.7, .8], in the interval [.77, .78], in the interval [.775, .776], etc., where the brackets [ ] indicate that the endpoints of the intervals are included. That is, n is in *every* one of the intervals determined in such a process; each of the intervals is inside the preceding interval; and each interval is $\frac{1}{10}$ as long as the preceding interval.

A number n with these characteristics is called a *real number*. A real number lies in every one of an unending sequence of nested intervals (that is, intervals within intervals, like a nested set of tables), with the ends of the intervals rational numbers and each interval $\frac{1}{10}$ the length of the preceding interval. Later in this section it will be shown that only one real number is in every interval of such a nest of intervals, and thus that each such nest of intervals defines exactly one real number. A real number so defined might coincide with an endpoint of some interval and be rational; it would then also be a real number because it is in every interval in the nest of intervals. On the other hand, it might be interior to every interval in the nest; then it is a real number that possibly is not rational. In either case, the number is a real number. If it is not rational, it is called *irrational*.

With the invention of real numbers we can say that with respect to a given unit, for each segment in the plane there is one real number that measures the segment, and if two segments are not congruent they have different measures. We can state this in terms of an infinite ruler (Fig. 2–30). For each point P on the edge of the ruler there is exactly one real number n, and for each real number n there is exactly one point P. Of course, this ruler, which associates with each point of the line a real number, is what is called the *number line*.

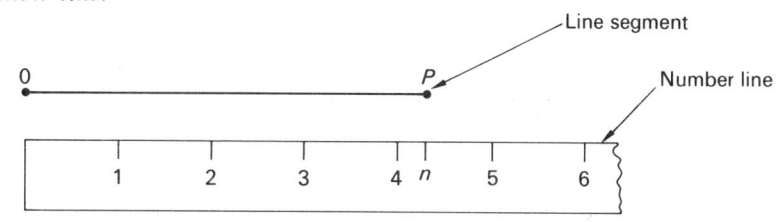

Figure 2–30

In elementary school mathematics programs, the concept of a real number is usually associated with that of an unending decimal. (The word decimal usually implies a terminating decimal, or one which "ends.") The notion of a real number, as developed here through the measuring process, fits the concept of an unending decimal. For example, the number n that is the measure of the segment $\overline{AB}$ in the example on page 55 can be represented as the unending decimal .7752 . . . If the number coincides with an endpoint of some interval of the nest, its decimal will contain an endless string of zeros. Thus every real number has an unending decimal representation. Conversely, every unending decimal corresponds to one real number that is located on the number line according to our measuring process.

For example, there is a positive real number whose square is 2 (in Section 3–7 we will construct a segment with this measure). To show an unending decimal for this number, again imagine successive magnifications of a ruler. (Denote by $\sqrt{2}$ a positive number whose square is 2. If this number obeys at least the properties of the rational numbers, then there can be *at most* one positive number whose square is 2. See Exercise 4 in Exercise Set 2–2.) Then, by *trial and error*, we find that

$$1 \leq \sqrt{2} \leq 2, \quad \text{since} \quad 1^2 \leq 2 \leq 2^2$$
$$1.4 \leq \sqrt{2} \leq 1.5, \quad \text{since} \quad (1.4)^2 \leq 2 \leq (1.5)^2$$
$$1.41 \leq \sqrt{2} \leq 1.42, \quad \text{since} \quad (1.41)^2 \leq 2 \leq (1.42)^2$$
$$\vdots$$

We see that $\sqrt{2}$, determined by this process, is in every interval, that the intervals are nested, and that each interval is $\frac{1}{10}$ the length of the preceding one. Thus $\sqrt{2}$ is a number defined by this nest of intervals and $\sqrt{2}$ is a real number represented by the unending decimal 1.414014 . . . Since there is at most one positive number whose square is 2, and since there is exactly one real number represented by each unending decimal (as we shall see at the end of this section), we decide that $\sqrt{2}$ is the real number represented by 1.414014 . . . (Later, in Chapter 3, we shall show that $\sqrt{2}$ is irrational.)

By successive divisions we see that $\frac{13}{8}$ is also a real number:

$$1 \leq \tfrac{13}{8} \leq 2$$
$$1.6 \leq \tfrac{13}{8} \leq 1.7$$
$$1.62 \leq \tfrac{13}{8} \leq 1.63$$
$$1.625 \leq \tfrac{13}{8} \leq 1.626$$
$$1.6250 \leq \tfrac{13}{8} \leq 1.6251$$
$$1.62500 \leq \tfrac{13}{8} \leq 1.62501$$
$$\vdots$$

The unending decimal representation of $\frac{13}{8}$ is 1.62500 . . . , where the zeros continue without end. This could be written as the decimal 1.625. Of course, $\frac{13}{8}$ is a rational number.

As a final example, consider this nest of intervals:

$$0 \leq \tfrac{2}{3} \leq 1$$
$$.6 \leq \tfrac{2}{3} \leq .7$$
$$.66 \leq \tfrac{2}{3} \leq .67$$
$$.666 \leq \tfrac{2}{3} \leq .667$$
$$\vdots$$

This shows that $\frac{2}{3}$ is a real number whose decimal .666 . . . is unending, with 6 repeated without end; $\frac{2}{3}$ is also a rational number.

It remains to consider this question: We know that every unending decimal represents a real number (because an unending decimal defines an unending nest of intervals), but is this representation unique? For example, is it possible that two different real numbers might be represented by the same unending decimal? The number line helps us decide that the representation is unique. Rephrase the question as follows: Given the unending nest of intervals defined by the unending decimal, are there two different points in every interval of the nest?

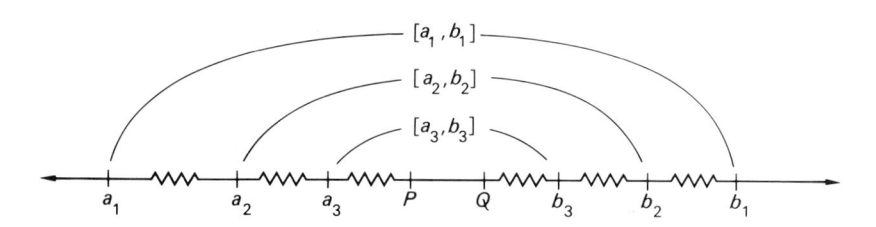

Assume that P and Q are two such points each lying in every interval of the nest. Then the line segment $\overline{PQ}$ has a fixed positive length. As the intervals, each $\frac{1}{10}$ the length of the preceding one, become smaller in length, surely some interval and all succeeding ones must be shorter than the length of $\overline{PQ}$. But then P and Q cannot both be in *every* interval of the nest. Thus we must conclude that at most one point is in every interval of the nest; correspondingly, at most one real number is represented by each unending decimal.

exercise set 2-2

1. Measure the line segment $\overline{AB}$ by giving at least three intervals in a set of nested intervals for its inch-measure, and present an argument that l, the inch-measure of $\overline{AB}$, is a real number.

2. Assume that there is a positive real number whose square is 3. Find the first three intervals in a set of nested intervals defining $\sqrt{3}$.

3. a) Show that $\frac{7}{4}$ is a real number. Find the first four intervals in a set of nested intervals defining $\frac{7}{4}$. [*Hint:* Begin with an interval whose endpoints are integers.]
   b) Show that $\frac{5}{12}$ is a real number. Find the first four intervals in a set of nested intervals defining $\frac{5}{12}$.

4. Suppose that there is a positive real number whose square is 2, and suppose that real numbers obey at least the properties of the rational numbers. Show that there is only one positive real number whose square is 2. (Assume that both a and b are positive numbers whose squares are 2. Then show that $a^2 - b^2 = 0$, so that

$$(a + b)(a - b) = 0.$$

Then show that $a = b$.)

teaching questions and projects 2–2

1. How would you explain to fifth graders the idea that physical measurement is approximate? Plan an activity to develop the meaning of "possible error" and "round-off error." If measuring instruments could be made sufficiently refined, is there ever the possibility that a physical measurement is a "true" measurement? Why?

2. Plan a discussion with children about the circumstances that determine how much error in measurement we wish to permit. For example, would measuring to the nearest $\frac{1}{4}$ inch be appropriate in all situations? In any situation?

3. Name some measurement situations in which the possible error could be one inch or more. Name some measurement situations in which the possible error could be $\frac{1}{16}$ inch or less.

4. Do you think elementary school children should be introduced to nonrational numbers? Defend your answer. If your answer is yes, at what level and to what extent? What background knowledge is prerequisite to the understanding of the real numbers?

5. Develop a lesson relating the idea of the nonrational number π to activities in which children determine the approximate measurements of the circumferences and diameters of many circles.

6. Develop a lesson in which children find $\sqrt{2}$ to the thousandth place by experimenting with squaring various numbers. (Do not use an algorithm.) For example, $\sqrt{2}$ must be between 1 and 2 because $1 \times 1 = 1$ and $2 \times 2 = 4$. Relate the activity to the concept of nested intervals.

7. Repeat Exercise 6 with other nonrational numbers, and then compare the decimal representation of some rational and some nonrational numbers.

chapter three

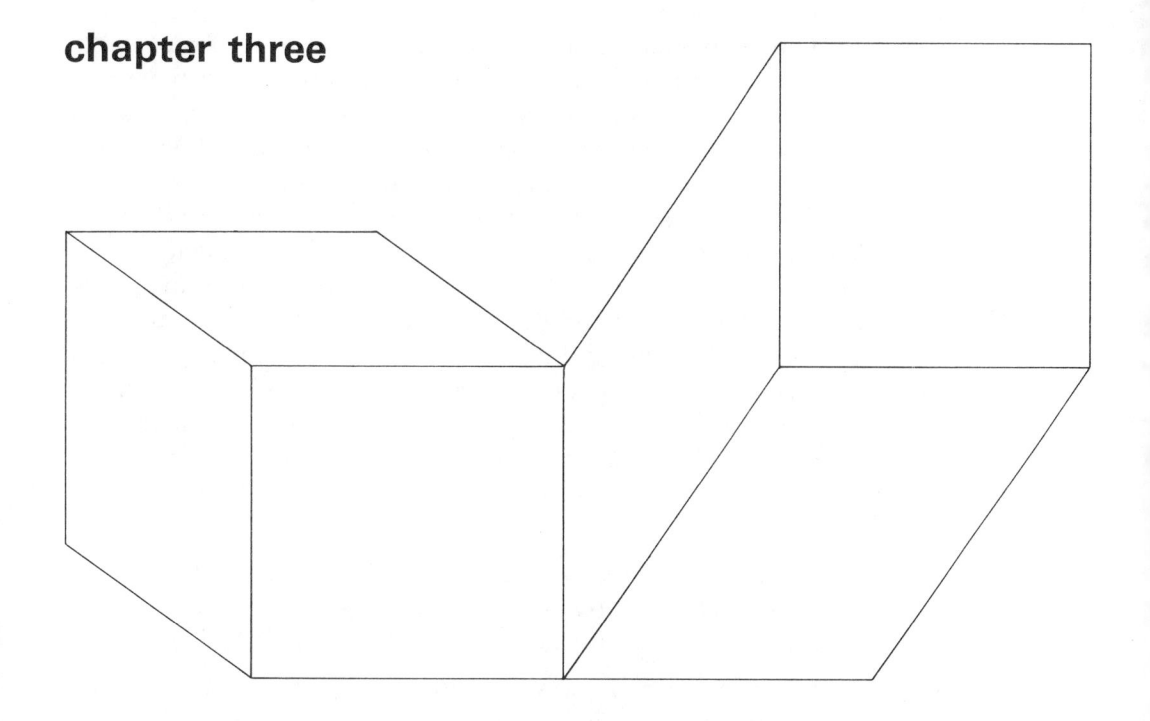

deductive geometry and constructions

3–1
introduction

Young children use familiar language to describe the geometric figures that are suggested by objects around them. At first, the physical objects are not distinct from the geometric figures suggested in their minds; to a child a bicycle tire *is* a circle.

As the distinction between a geometric figure (an abstraction) and an object suggesting the figure becomes clearer, the language used to describe this distinction needs to be clarified also. (One of the criticisms of traditional courses in geometry has to do with the fuzziness of such language.)

Thus elementary school children begin to sense, for example, that a dot of ink on a paper may represent a point, but the notion of a point is suggested by an exact position in space and cannot be perfectly represented by any dot, no matter how fine. They also begin to sense that among the many properties of figures that are discovered, some are consequences of others and can be determined directly from the others without experimenting with the figures. For example, after they measure many triangles and find that the sum of the degree measures of the three angles of each triangle is always 180, they can use this property, without measuring more triangles, to conclude that if two angles of one triangle are congruent to two angles of another triangle, then the third angles of the two triangles are also congruent. These and other realizations suggest new levels of activity in the study of geometry:

1. We need to develop language that will support the notions of geometric figures as idealizations of objects (and that at the same time will be a natural and uncomplicated language).

2. Since geometric figures are abstract concepts, we cannot place too much reliance on the rough figures we draw when we make decisions about the properties of figures; we therefore need to learn a strategy of showing (proving) that certain properties are true in all cases regardless of our intuition or our rough drawings.

3. We need to identify a small set of properties that are basic and then show that all others are consequences of these.

In this chapter we shall begin a consideration of these three levels of activity. The first—the development of language—will involve the introduction and use of the language of sets and their unions and intersections in order to give careful descriptions of geometric figures. We shall begin the second—the development of strategies for showing certain properties to be true in all cases—by considering some geometric constructions and showing that they hold true generally. In the process we shall review the standard constructions and at the same time illustrate methods of proof. The third—the rigorous deduction of all of plane geometry from a set of assumed basic properties—is a project too ambitious for an elementary program. Indeed, this was done satisfactorily for the first time less than a century ago and is still a rather sophisticated exercise in mathematics. Instead we shall show the need for some systematization in studying geometry. By means of a list of assumed properties, though incomplete, we can learn how some parts of plane geometry can be formally deduced from these properties.

Our main emphasis will be on learning some ideas about deduction in general, and in particular on justifying some constructions of plane geometry. First, appropriate language will be developed, and then the idea of a geometric system will be illustrated.

3-2
space as a set of points

Let us assume that the reader is familiar with the simple language of sets, particularly with respect to sets of numbers in the beginning notions of arithmetic. It will then be natural to use the same language in talking about sets of points.

For example, without trying to define point, we shall agree that *space is the set of all points*. Thus every geometric figure that we discussed intuitively in Chapters 1 and 2 can be regarded as a subset of space, that is, as a set of points in space. For example, a line is a set of points in space; in Chapter 2 we tacitly used this description when we assigned to each point of a line one and only one real number to form what is called an infinite ruler or the number line.

The language of sets will evolve quite naturally as we proceed. In the following paragraphs we include some examples of the use of this language in describing our intuitive geometric concepts; other examples can be provided by the reader.

Given two points A and B, we agreed there is exactly one line $\overleftrightarrow{AB}$ through A and B. That is, the line $\overleftrightarrow{AB}$ is a set of points, two of which are A and B. In Chapter 1 we said that point A *is on* the line $\overleftrightarrow{AB}$ and that the line $\overleftrightarrow{AB}$ *passes through* point A. Now we can say that point A *belongs to* line $\overleftrightarrow{AB}$ (in the sense of an element belonging to a set), and line $\overleftrightarrow{AB}$ *contains* point A. The line segment $\overline{AB}$ can then be described as the *subset* of $\overleftrightarrow{AB}$ consisting of A and B (called the endpoints of $\overline{AB}$) and all the points of $\overleftrightarrow{AB}$ between A and B. (In general, a set T is a subset of a set Y

64 deductive geometry and constructions

if every element belonging to T also belongs to Y. If T is a subset of Y we say that T is *contained in Y*.)

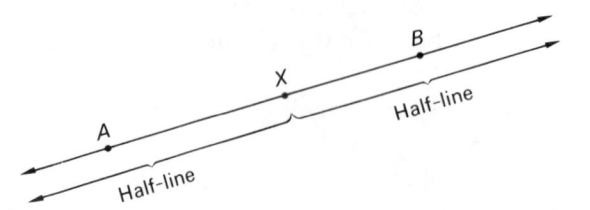

Figure 3–1

Any point belonging to $\overleftrightarrow{AB}$ separates $\overleftrightarrow{AB}$ into two half-lines. In Fig. 3–1, where X happens to be between A and B, the point B belongs to one half-line and A belongs to the other half-line. The point X is in neither of its half-lines. Thus the ray $\overrightarrow{XB}$ can be described as the set of points consisting of X (the endpoint of the ray) and all the points in the half-line on the B-side of X.

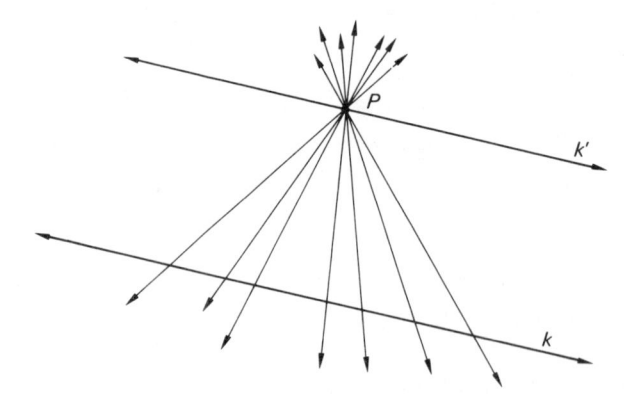

Figure 3–2

We agreed in Chapter 1 that a line k together with a point P not on k determine a plane. The set of points making up this plane can now be described in the following way, provided we assume (as we shall in Section 3–5) that there is exactly one line through P parallel to k. In Fig. 3–2, for each point in k we have a unique line determined by this point and P. This gives us a set of lines each containing P and each having all its points in the plane. Note that there is another line in the plane containing P that has not been included; it is the line k' parallel to k through P. Finally, if we take the set of all the lines determined by P and points on k together with the line k' containing P and parallel to k, we have what is called a *plane pencil* of lines through P. The plane determined by P and k is the set of all points in space belonging to lines of this pencil.

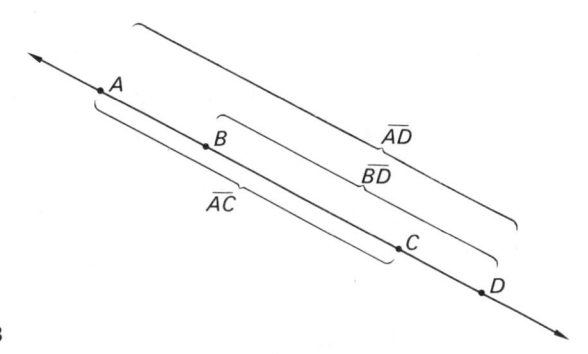

Figure 3–3

The language of sets can further clarify some of the preceding discussions. For example, a ray can be described in terms of the union of a point and a half-line; a point at which two lines meet can be described in terms of the intersection of the lines. The words union and intersection have the expected meanings. (If S and T are two sets, then the *union* of S and T $(S \cup T)$ is a set consisting of all elements either in S or in T; the *intersection* of S and T $(S \cap T)$ is a set consisting of all elements in both S and T.) For example, in Fig. 3–3,

$$\overline{AC} \cap \overline{BD} = \overline{BC}$$

because each point of $\overline{BC}$ is in both $\overline{AC}$ and $\overline{BC}$. In the same figure we see that

$$\overline{AC} \cup \overline{BD} = \overline{AD}$$

because each point of $\overline{AD}$ is either in $\overline{AC}$ or in $\overline{BD}$, that is, in at least one of $\overline{AC}, \overline{BD}$. In the same figure,

$$\overleftrightarrow{AB} \cup \overline{BC} = \overleftrightarrow{AB},$$

$$\overleftrightarrow{AB} \cap \overline{BC} = \overline{BC},$$

and

$$\overline{AB} \cap \overline{CD} = \emptyset.$$

Since there are no points that are in both $\overline{AB}$ and $\overline{CD}$, we say that their intersection is empty. (The symbol $\emptyset$ denotes the empty set, that is, the set containing no members.) Other relations given by the same figure are:

$\overrightarrow{AB} \cup \overrightarrow{BA} = \overleftrightarrow{AB}$ (since each point of line $\overleftrightarrow{AB}$ is in at least one of the rays $\overrightarrow{AB}, \overrightarrow{BA}$)
$\overrightarrow{AB} \cap \overrightarrow{BA} = \overline{AB}$ (since each point of $\overline{AB}$ is in both rays)
$\overrightarrow{AC} \cup \overrightarrow{BD} = \overrightarrow{AD}$ (Why?)
$\overrightarrow{AC} \cap \overrightarrow{BD} = \overline{BD}$ (Why?)
$\overrightarrow{AB} \cap \overrightarrow{BC} = \{B\}$ (since B is the only point in both $\overrightarrow{AB}$ and $\overrightarrow{CB}$)
$\overrightarrow{BA} \cap \overrightarrow{CD} = ?$
$\overrightarrow{CB} \cap \overrightarrow{CD} = ?$

We can describe other familiar geometric figures in a plane in the same way. An angle, for example, is a set of points; it is the union of two rays with a common endpoint: $\angle ABC = \overrightarrow{BA} \cup \overrightarrow{BC}$. (Explain how this description allows the possibility of 0° angles and 180° angles.) A triangle, for example, is a union of three segments: $\triangle ABC = \overline{AB} \cup \overline{BC} \cup \overline{CA}$. What is a description of a quadrilateral?

Curves in a plane are also examples of sets of points. A simple closed curve determines a *region* in the plane, which is a subset of the plane that consists of the union of the simple closed curve and the set of points inside the curve.

As another example of the use of set language in geometry, consider the lines, the triangle, and the region in Fig. 3–4.

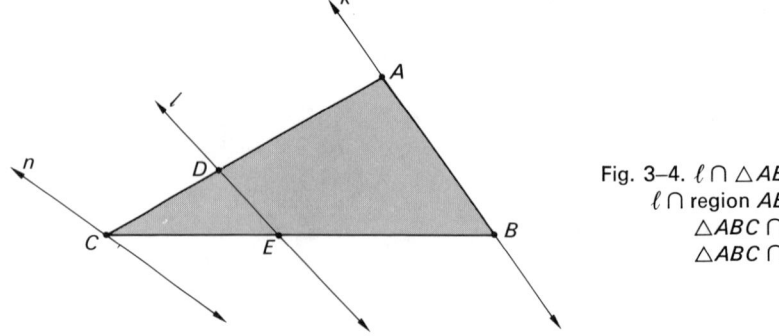

Fig. 3–4. $\ell \cap \triangle ABC = \{D, E\}$,
$\ell \cap$ region $ABC = \overline{DE}$,
$\triangle ABC \cap n = \{C\}$,
$\triangle ABC \cap k = \overline{AB}$.

Each plane is a certain set of points in space. Any line l contained in plane $\mathbb{P}$ separates $\mathbb{P}$ into two half-planes. In Fig. 3–5, point Q belongs to one of the half-planes of $\mathbb{P}$ determined by l and

R belongs to the other half-plane. The line l is in neither of its half-planes. Note that the segment $\overline{QR}$ intersects l. What can you say about the intersection of $\overline{AB}$ and l if both A and B belong to the same half-plane of $\mathbb{P}$ formed by l?

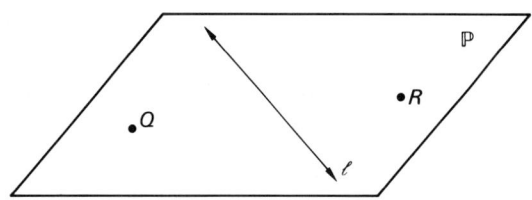

Figure 3–5

The points inside an angle (other than a 0° or 180° angle) can be described as an intersection as follows: Given angle XYZ as shown in Fig. 3–6, consider the line $\overleftrightarrow{XY}$ and denote by $\mathbb{B}$ the half-plane formed by $\overleftrightarrow{XY}$ that contains the point Z. Let $\mathbb{A}$ be the half-plane formed by $\overleftrightarrow{YZ}$ that contains the point X. Then the set of points inside $\angle XYZ$ is $\mathbb{A} \cap \mathbb{B}$.

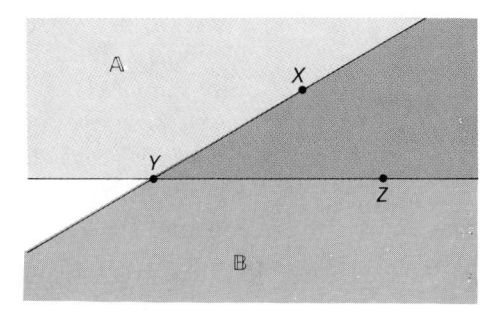

Figure 3–6

3–3
assumptions about sets of points in a plane

Definitions In the preceding section we "described" some figures as sets of points. We already had an idea of each figure from our early mathematical experiences with physical models, and we merely stated that idea in the language of sets of points.

In the later years of elementary school, children may devise simple proofs about properties of figures that hold true generally. They are then ready to study some deductive geometry. Before they can do this, there must be agreement as to the meanings of the terms used in the proofs. Although "descriptions" were satisfactory for the purposes of early mathematical experiences, *definitions* are needed now.

Since a definition must give a meaning to what is being described, the question arises: Is it possible to define anything? If so, we must understand the meaning of each word used in the definition. But this requires that we give a definition of each word used. The process could go on forever, each time using new words that still need to be defined. Clearly, in geometry as well as in other areas we must stop with certain basic words and agree that we accept their meanings without definition. We take certain terms as *undefined*, and we define other words using these undefined terms.

As would be suspected, in our approach to deductive geometry the word point will be left undefined. (We find some unsuccessful attempts in some traditional geometry books to define point, in which the result is a description in terms of other words even more difficult to define.) Some other undefined terms will be: line, plane, length, measure, congruent, between (in the sense that a point is between two given points or a ray is between two given rays). We shall also accept set language without analyzing its basic terms, which include: element, set, belong, union, intersection.

Using these undefined, or primitive, terms we can define other geometric terms. For example, consider this definition.

▶ *Definition 1*

a) *Space* is the set of all points.

b) A *geometric figure* is a nonempty set of points in space.

(Note that the definition uses the undefined term point and the set language set and nonempty set. We can define *nonempty set* as a set to which at least one element belongs, thereby reducing our basic set language to the undefined words set, element, and belong.)

postulates and theorems As we continue to define geometric words by use of the undefined words—point, line, plane, length, and between—we shall have in mind the meanings of these undefined words suggested by our physical experiences. Thus, although we do not define points, lines, and planes in deductive geometry, we do have some idea of the properties that we expect them to possess. Whatever points and lines are, we would require, for example, that exactly one line contains two given points.

This procedure of giving meaning to basic terms will be one of characterization rather than definition. (We do this also in nongeometric contexts. For example, although we would hesitate to define love, we could characterize love by describing its manifestation and the effects it has on its recipients.)

In a similar way we shall give geometric meaning to the words point, line, and plane by making statements about their incidence properties (intersection properties) that we agree or assume to be true of all points, lines, and planes, no matter what these objects are. Such assumed statements are called *postulates*. (The undefined word *between* can also be described in a way fitting our intuition by listing properties that points or rays should obey. We shall not do this here; instead we accept our intuitive meaning of "between" with the understanding that it could be postulated in a more rigorous development.)

We also want to be able to measure geometric figures, and this requires that we characterize the undefined word *measure* by listing properties that it should satisfy generally.

There are other properties of figures that are suggested by our physical experiences, but which we cannot prove as consequences of the postulates already assumed. These will also have to be assumed and listed as postulates. Finally, it should be possible to prove all other facts in geometry as consequences of this set of postulates. The consequences are called *theorems*.

Each branch of mathematics could be developed deductively in this way. We accumulate wide intuitive experiences with objects in a physical setting; then we systematically organize the experiences and from them abstract general concepts about the objects or properties of the objects. We take certain statements about the concepts that characterize them and accept these statements as postulates; then we prove as many consequences (theorems) as we can, some of which might give us new insights about the physical setting that we never recognized before. In the process, we use certain words, some of which are left undefined and are characterized by the postulates; other words are defined in terms of the undefined words.

Thus a formal deductive system consists of a collection of undefined terms, a set of postulates concerning these terms, definitions of new terms, and the logical consequences or theorems deduced from the postulates and definitions.

We shall not try to set up a complete deductive system of geometry, but we shall give some of the postulates and definitions of a space geometry and show how some theorems in plane geometry are consequences. Our object is to be able to prove some of the usual geometric constructions in a plane, and in the process to follow a partial deductive development of plane geometry, with some of the details occasionally omitted.

proof Before proceeding to the partial development of a geometric system, we include a few words about proof, its nature and method.

A deductive proof in a formal sense is a series of sentences, each following from the ones before by logical rules. We might call the proof an argument, in which we begin with some assumed statements and then see what other statements can be derived logically from them. The conclusion of the argument is valid if it follows as a logical consequence of the assumptions. Thus the conclusion of an argument is valid by virtue of its form, regardless of the truth or falsity of the assumed statements. It is important to note, however, that if the statements from which we derive a valid conclusion are true, then the conclusion is also true.

The proofs in this chapter are informal and rely on the reader's intuitive understanding of what is reasonable and what is valid. Most often they proceed by appealing to a postulate or several postulates (or perhaps to definitions) and say, "If Postulate such-and-such is true, then it follows that . . ." or "By Postulate such-and-such, we see that . . ."

Perhaps the key phrase in any proof is the phrase "if . . . , then . . ." In general, the reader will be determining in a proof what logically follows if we accept the postulates, definitions, and previously proved theorems. Or, given the postulates, definitions, and previously proved theorems as true statements, he will be determining what seems to be reasonable and logical consequences of those statements and why.

first postulates and definitions

Postulate 1. (Incidence)

a) Lines and planes are sets of points in space.

b) There are at least two points belonging to each line. Given two points, there is exactly one line containing the points.

c) There are at least three points, not all in one line, belonging to each plane. Given any three points, not all in one line, there is exactly one plane containing the points.

d) There are at least four points, not all in one plane, belonging to space.

e) If two points belong to a plane, then the line containing the points is contained in the plane.

f) If two planes intersect, their intersection is a line.

With the assumption (Postulate 1b) that each pair of points determines exactly one line, we can make the following definitions.

▶ *Definition 2.* A *line segment* $\overline{AB}$ is a set consisting of points A and B and all the points of line $\overleftrightarrow{AB}$ that are between A and B. (We call A and B the *endpoints* of $\overline{AB}$.)

▶ *Definition 3.* A *ray* $\overrightarrow{AB}$ is the union of $\overline{AB}$ and the set of all points X belonging to $\overleftrightarrow{AB}$ such that B is between A and X. (A is called the *endpoint* of ray $\overrightarrow{AB}$.)

▶ *Definition 4.* An *angle* $\angle CAB$ is the union of ray $\overrightarrow{AC}$ and ray $\overrightarrow{AB}$. (The point A is called the *vertex* of $\angle CAB$, and $\overrightarrow{AC}$ and $\overrightarrow{AB}$ are called its *sides*.)

The terms triangle, quadrilateral, polygon, and so on must be similarly defined. (See Exercise Set 3–1.)

The reader will recall some other results about incidence of lines that are listed in Chapter 1. These should now be consequences of Postulate 1. For example, we can prove that two lines intersect in at most one point. The proof is as follows: Suppose that lines l and k intersected in two points, P and Q. Then both P and Q would belong to l and both would belong to k. But Postulate 1(b) requires that there be exactly one line containing P and Q, not two. Thus l and k cannot intersect in more than one point.

As another example of a consequence of the incidence postulate we can prove that, given two points A and B, there is exactly one line segment with these points as endpoints. We leave the argument or proof of this result to the reader.

Postulate 2. (Measurement of segments)

a) Given a segment (called the *unit* segment), then corresponding to each segment $\overline{AB}$ there is a positive real number called the *measure* of $\overline{AB}$ and denoted by AB; conversely, given any ray $\overrightarrow{AC}$ and any positive real number t, there is a point B on $\overrightarrow{AC}$ such that the measure of $\overline{AB}$ is t. The measure of the unit segment is 1, and two segments are congruent if and only if they have the same measure with respect to the unit segment. (The *length* of $\overline{AB}$ or the *distance between* A and B is the pair consisting of the measure of $\overline{AB}$ and the unit segment.)

b) If X (Fig. 3–7) belongs to the segment $\overline{AB}$, then

$$AX + XB = AB.$$

c) If X does not belong to $\overline{AB}$, then $AB < AX + XB$.

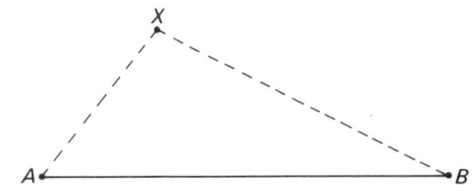

Figure 3–7

Using the notion of measure we can now define a circle.

▶ **Definition 5.** A *circle* with the point C as *center* is a figure consisting of all points X in a plane containing C such that $CX = r$, where r is a positive real number. The length of $\overline{CX}$ for any point X on the circle is called the *radius* of the circle.

(Sometimes we speak loosely of the circle with center C and radius r, meaning that the radius is r units.)

We make a similar characterization of measurement of angles.

Postulate 3. (Measurement of angles)

a) Given an angle, called a *unit* angle, then corresponding to each ∠ABC there is a positive real number, denoted by

$m(\angle ABC)$, called the *measure* of $\angle ABC$; and conversely, given any ray $\overrightarrow{AB}$ and given any real number t, there is a ray $\overrightarrow{AC}$ such that $m(\angle BAC) = t$. Two angles are *congruent* if and only if they have the same measure, and the measure of the unit angle is 1.

(Note that in Postulates 2(a) and 3(a) we defined "congruent segments" and "angles" in terms of the undefined word measure. In a more formal course the word congruent would be left undefined and characterized by a set of postulates. In general, we shall take the meaning of congruence of figures to be as described in Chapter 1 in terms of rigid motions.)

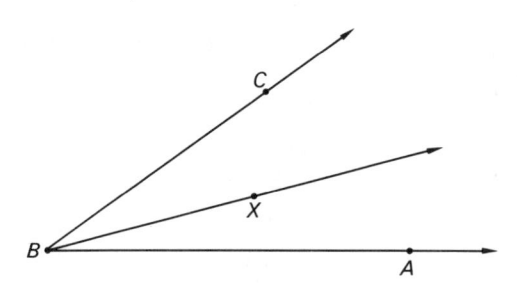

Figure 3–8

b) If $\overrightarrow{BX}$ (Fig. 3–8) is between $\overrightarrow{BA}$ and $\overrightarrow{BC}$ (that is, if X is inside $\angle ABC$), then $m(\angle ABX) + m(\angle XBC) = m(\angle ABC)$.

c) The measure of a straight angle (an angle whose rays form a line) in degrees is 180.

For some of our work on geometric constructions we also need the following definitions.

▶ **Definition 6.** Two angles, $\angle BAD$ and $\angle DAC$, that are formed by the ray $\overrightarrow{AD}$ and the line containing B, A, and C are called a *linear pair* of angles (Fig. 3–9).

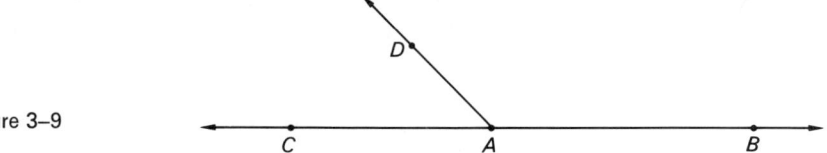

Figure 3–9

▶ **Definition 7.** When two angles of a linear pair are congruent, they are called *right angles*.

It follows immediately that the sum of the degree measures of two angles of a linear pair is 180, and that the degree measure of every right angle is 90. (Why?) Any two angles whose degree measures have the sum 180 are called *supplementary angles*.

construction of congruent line segments Construction problems in geometry have to do with constructing figures that are congruent to given figures. Recall that we are assuming that the term congruent figures has the general meaning developed in Chapter 1 and the particular meaning of congruent segments and angles given in Postulates 2 and 3.

We want to have "enough" points on lines and circles so that intersecting lines and circles will have points in common. To guarantee enough points we list another postulate.

Postulate 4. (Incidence of lines and circles)

a) Every ray which is in the plane of a circle and whose endpoint is at the center of the circle intersects the circle in exactly one point.

b) If C_1 and C_2 are circles in the same plane with centers at A and B and with radii a and b, respectively, and if each of the numbers a, b, c ($c = AB$) is less than the sum of the other two, then C_1 and C_2 intersect in exactly two points X and Y, and $\overline{XY}$ intersects $\overline{AB}$ in a point between A and B (Fig. 3–10).

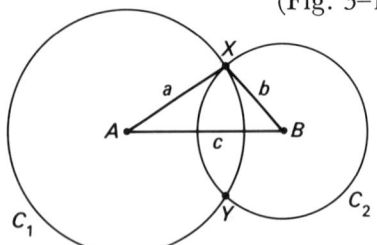

Figure 3–10

c) If $\overleftrightarrow{AB}$ is a line in a plane $\mathbb{P}$ and if X is any point on $\mathbb{P}$, then there is a circle with center at X that intersects $\overleftrightarrow{AB}$ in exactly two points.

We begin with the most elementary geometric construction.

• *Problem 1.* Given a line segment $\overline{AB}$ and a ray $\overrightarrow{XY}$, construct a line segment $\overline{XZ}$ with Z in $\overrightarrow{XY}$ so that $\overline{XZ}$ is congruent to $\overline{AB}$.

Construction 1. Set a compass to have radius AB. Then draw a circle with center at X. Let Z be the point in $\overrightarrow{XY}$ at which the circle intersects $\overrightarrow{XY}$. Then $\overline{XZ}$ is the required segment (Fig. 3–11).

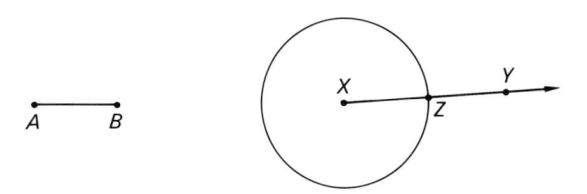

Figure 3–11

For specific cases the above construction can be seen to produce a segment congruent to a given segment. Will it provide the required segment in *every* case? Let us prove that it will.

Proof of Construction 1. For any segment $\overline{AB}$ there is a positive number r measuring $\overline{AB}$, by Postulate 2(a). Then for any ray $\overrightarrow{XY}$ a circle with center at X and radius r will intersect $\overrightarrow{XY}$ in one point, say Z, such that $XZ = r$. Thus AB and XZ have the same length and are therefore congruent. We have proved the construction for all cases.

construction of congruent angles An analogous construction can be described for copying a given angle.

• *Problem 2.* Given an angle $\angle ABC$ and a ray $\overrightarrow{XY}$ in a plane, construct two points D and E in the plane such that D belongs to $\overrightarrow{XY}$ and $\angle DXE$ is congruent to $\angle ABC$.

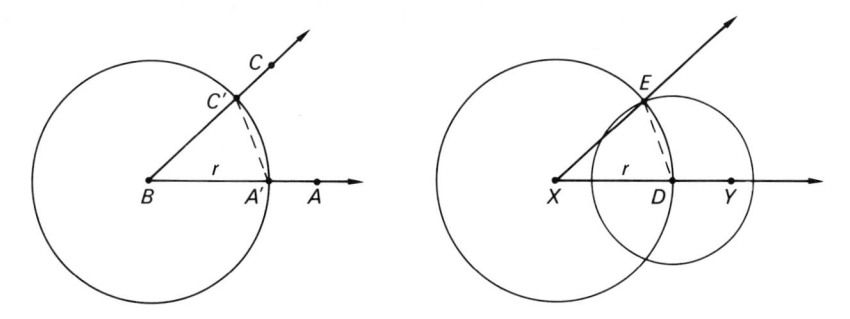

Figure 3–12

Construction 2. Set a compass to have any radius r, and draw a circle with center at B (Fig. 3–12). Let C' and A' be the points of $\overrightarrow{BC}$ and $\overrightarrow{BA}$ at which the circle intersects these rays, respectively. With the same radius r, draw a circle with center at X. Call D the intersection of this circle with $\overrightarrow{XY}$. Then with the compass set to have radius $A'C'$, draw a circle with center at D. By Postulate 4(b), the circles with centers at X and D intersect in two points. (Why? What is true of the three numbers r, $A'C'$, XD?) Call one of the points E. Then $\angle DXE$ is the required angle congruent to $\angle ABC$.

Again we can verify that this construction works in specific cases, and again we hope to prove that it will work in all cases. Note that in the construction shown in Fig. 3–12 there are two triangles formed, $\triangle BC'A'$ and $\triangle XED$. The proof will depend on showing that these two triangles are congruent, and therefore that $\angle EXD$ is congruent to $\angle C'BA'$. This suggests that before we can prove Construction 2 we need to discuss congruent triangles and state some postulates concerning them.

3–4
assumptions about congruent triangles

We described two plane figures as congruent (Chapter 1) if one figure could be moved onto the other by a series of parallel, turning, or folding movements, and agreed that these movements preserved congruence of all parts of the figure. Thus, in order to describe congruence of two figures, we must describe where points go when one figure is moved onto the other. For example, if triangle ABC (Fig. 3–13) is moved onto triangle DEF so that A goes to D, B to E, C to F, we have a matching of vertices called a *one-to-one correspondence* between the vertices of the triangles that can be written $A \leftrightarrow D$, $B \leftrightarrow E$, $C \leftrightarrow F$ or $ABC \leftrightarrow DEF$. (Of course, there are various correspondences between the vertices of two given triangles. For example, the correspondence $BAC \leftrightarrow DFE$ matches vertex B with vertex D, A with F, and C with E.) For

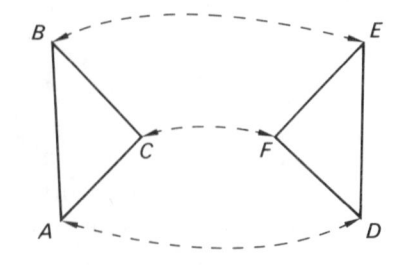

$ABC \leftrightarrow DEF$

Figure 3–13

congruent triangles we make the following definition (here the symbol ≡ means "is congruent to"):

▶ *Definition 8.* If $ABC \leftrightarrow DEF$ is a correspondence between the vertices of two triangles such that $\overline{AB}$ is congruent to $\overline{DE}$ ($\overline{AB} \equiv \overline{DE}$), $\overline{BC} \equiv \overline{EF}$, $\overline{CA} \equiv \overline{FD}$, and such that $\angle ABC$ is congruent to $\angle DEF$ ($\angle ABC \equiv \angle DEF$), $\angle BCA \equiv \angle EFD$, $\angle CAB \equiv \angle FDE$, then the correspondence $ABC \leftrightarrow DEF$ is a *congruence* between the two triangles ($\triangle ABC \equiv \triangle DEF$) and conversely. (Note that we are careful to write the vertices in the order in which they are paired.)

In effect, Definition 8 says that two triangles are congruent if and only if there is a correspondence between their vertices under which each of their pairs of corresponding sides and angles are separately congruent. Which parts of one triangle will need to be congruent to the corresponding parts of another triangle to guarantee that the two triangles are congruent? Our intuitive experiences lead us to agree that the congruence of all sides of one triangle with the corresponding sides of the other is sufficient for congruence of the triangles. We make this assumption and list it as a postulate.

Postulate 5. (Congruence of triangles)

a) If there is a correspondence between the vertices of two triangles such that each side of one triangle is congruent to the corresponding side of the other, then the triangles are congruent.

With this postulate (referred to as the side-side-side or SSS postulate) we can prove that the construction in Problem 2 will hold in all cases.

Proof of Construction 2. (See Fig. 3–12.)
For *any* $\angle ABC$ and for *any* ray $\overrightarrow{XY}$, $\overline{BA'} \equiv \overline{XD}$ because $BA' = r$ and $XD = r$, by Postulate 2(a); $\overline{BC'} \equiv \overline{XE}$ because $BC' = r$ and $XE = r$, by Postulate 2(a); $\overline{A'C'} \equiv \overline{DE}$ because $A'C'$ and DE are equal radii; $\triangle A'BC' \equiv \triangle DXE$ because of the SSS postulate; $\angle A'BC' \equiv \angle DXE$ because corresponding angles of congruent triangles are congruent, by Definition 8.

Thus ∠*DXE* is the required angle. We have proved the construction for *all* cases.

In advanced geometry courses the following are proved as theorems. In an elementary program we usually take them as postulates.

Postulate 5

b) If there is a correspondence between the vertices of two triangles such that two sides and their included angle in one triangle are congruent to the corresponding two sides and included angle in the other, then the triangles are congruent. (This is the side-angle-side, or SAS postulate.)

c) If there is a correspondence between the vertices of two triangles such that two angles and their common side in one triangle are congruent to the corresponding two angles and their common side in the other, then the triangles are congruent. (This is the angle-side-angle, or ASA postulate.)

There are three sides and three angles in a triangle. In Postulates 4 and 5 we assume that if a certain three of these parts are congruent to the corresponding three of another triangle, then the triangles are congruent. The symbols SSS, SAS, and ASA help us to specify the cases we assume will yield congruence. We can quickly dispose of two of the remaining cases, AAA and SSA, by exhibiting noncongruent triangles with these congruent parts.

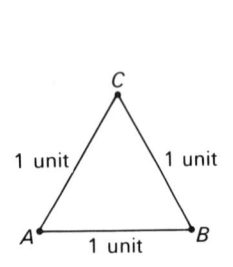

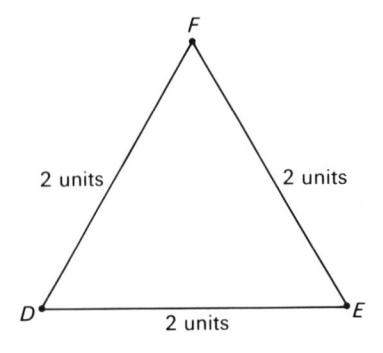

Figure 3–14

Case AAA: Consider the two equilateral triangles *ABC* and *DEF* (that is, triangles with each angle measuring 60°), with each side of △*ABC* 1 unit long and each side of △*DEF* 2 units long (Fig.

3–14). Then under *any* correspondence between the vertices of these two triangles, the corresponding angles are congruent. But under *none* of these correspondences are any corresponding sides congruent. Hence AAA does not yield congruent triangles.

(We shall postpone until Chapter 6 a discussion of the AAA case, but $\triangle ABC$ is said to be *similar* to $\triangle DEF$ if there is a positive real number k such that $DE = k(AB)$, $FE = k(CB)$, and $DF = k(AC)$.)

Case SSA: Choose the correspondence $ABC \leftrightarrow DEF$ (see Fig. 3–15). Then $\overline{BC} \equiv \overline{EF}$, $\overline{CA} \equiv \overline{FD}$, $\angle A \equiv \angle D$, but $\triangle ABC \not\equiv \triangle DEF$ because $\overline{AB} \not\equiv \overline{DE}$ ("$\not\equiv$" is read "is not congruent to"). This is the only correspondence between these triangles that yields SSA. Hence, SSA does not yield congruent triangles.

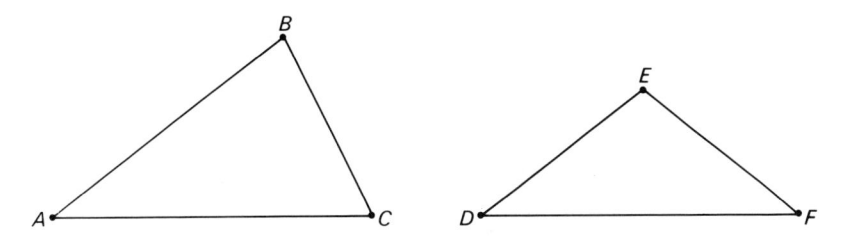

Figure 3–15

The remaining case, AAS, yields congruent triangles. The proof of AAS depends on a theorem that will be proved later (see Exercise 2, Exercise Set 3–2). For the present we will assume it.

The sum of the degree measures of the angles of any triangle is 180.

Let us accept this theorem for the moment and use it to prove Theorem 1 below. (As we proceed we must of course be careful that we do not use any consequences of the above theorem when we prove it later. We have accepted it in advance in order to get on with proofs of constructions that are of interest to us here.)

▶ *Theorem 1.* (AAS) If there is a correspondence between vertices of two triangles such that two angles and a side opposite one of the angles in one triangle are congruent to the corresponding two angles and side in the other, then the triangles are congruent.

Proof of Theorem 1. Let the triangles be $\triangle ABC$ and $\triangle DEF$, with the correspondence $ABC \leftrightarrow DEF$, so that $\angle A \equiv \angle D$, $\angle C \equiv \angle F$, and $\overline{AB} \equiv \overline{DE}$.

Then $m(\angle A) + m(\angle C) + m(\angle B) = 180$ and $m(\angle D) + m(\angle F) + m(\angle E) = 180$ because the sum of the degree measures of the angles of any triangle is 180.

Also $m(\angle B) = 180 - [m(\angle A) + m(\angle C)]$ and $m(\angle E) = 180 - [m(\angle D) + m(\angle F)]$ because of the properties of addition of real numbers.

Also $m(\angle A) + m(\angle C) = m(\angle D) + m(\angle F)$ because $m(\angle A) = m(\angle D)$ and $m(\angle C) = m(\angle F)$, and properties of real numbers.

Thus $m(\angle B) = m(\angle E)$ because of the property of equality. Then $\triangle ABC \equiv \triangle DEF$ because of Postulate 5(c) (ASA). The proof is now complete.

constructions of perpendicular lines Some interesting constructions involve constructing lines perpendicular to given lines.

▶ *Definition 9.* Two intersecting lines are *perpendicular* to each other if and only if an angle formed by the lines is a right angle. Two rays or segments are perpendicular to each other if and only if the lines containing them are perpendicular. (We shall use the symbol $\perp$ to mean "is perpendicular to.")

• *Problem 3.* Given $\overleftrightarrow{AB}$ and a point X in $\overleftrightarrow{AB}$ between A and B, construct $\overleftrightarrow{XY}$ so that $\overleftrightarrow{XY} \perp \overleftrightarrow{AB}$ (see Fig. 3–16).

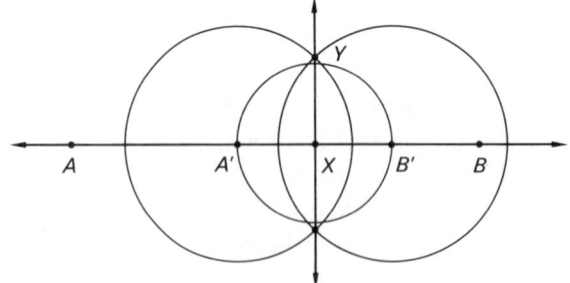

Figure 3–16

Construction 3. Set the compass with any radius r and draw a circle with center at X. By Postulate 4(a), this circle intersects $\overrightarrow{XA}$ at some point A' and intersects $\overrightarrow{XB}$ at some point B'. Set the compass with a radius s so that $s > r$. Then draw two circles, one with center at A' and the other with center at B'. These two circles intersect in two points (Postulate 4b), because each of the numbers s (radius of one circle), s (radius of other circle), and $2r$ (distance between their centers) is less than the sum of the other two; in particular, $2r < s + s$ because $r < s$.

Let one of the points of intersection of these circles be called Y. Then $\overleftrightarrow{XY} \perp \overleftrightarrow{AB}$.

Proof of Construction 3. For any $\overleftrightarrow{AB}$ and any X in $\overleftrightarrow{AB}$ between A and B, $\overline{A'X} \equiv \overline{B'X}$ because $A'X = r$ and $B'X = r$, and by definition of a circle these segments are congruent. $\overline{A'Y} \equiv \overline{B'Y}$ because $A'Y = s$ and $B'Y = s$, and by definition these segments are congruent. $\overline{XY} \equiv \overline{XY}$ because $\overline{XY} = \overline{XY}$.

Then $\triangle A'XY \equiv \triangle B'XY$ because of Postulate 5(a) (SSS). $\angle A'XY \equiv \angle B'XY$ because corresponding angles of congruent triangles are congruent. $\angle A'XY$ and $\angle B'XY$ are right angles because $\angle A'XY$ and $\angle B'XY$ are congruent angles of a linear pair (Definition 7); $\overleftrightarrow{XY} \perp \overleftrightarrow{AB}$ because an angle formed by $\overleftrightarrow{XY}$ and $\overleftrightarrow{AB}$ is a right angle, by Definition 9.

Thus $\overleftrightarrow{XY}$ is the required line through point X on $\overleftrightarrow{AB}$ and perpendicular to $\overleftrightarrow{AB}$, and we have proved the construction.

• *Problem 4.* Given $\overleftrightarrow{AB}$ and a point X not contained in $\overleftrightarrow{AB}$, construct $\overleftrightarrow{XY}$ such that $\overleftrightarrow{XY} \perp \overleftrightarrow{AB}$ (see Fig. 3–17).

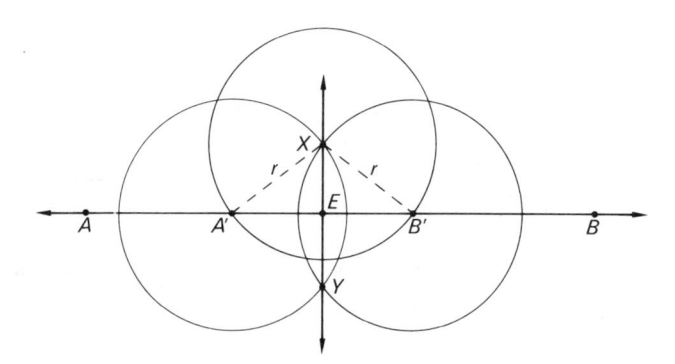

Figure 3–17

Construction 4. Set the compass with radius r large enough so that a circle drawn with center X (Fig. 3–17) intersects $\overleftrightarrow{AB}$ in two points A' and B', by Postulate 4(c). Then with the same radius r

draw two circles, one with center at A' and the other with center at B'. These circles intersect at X and at another point Y by Postulate 4(b), because each of the numbers r, r, and $A'B'$ is less than the sum of the other two. Then $\overleftrightarrow{XY} \perp \overleftrightarrow{AB}$.

Proof of Construction 4. The reader can provide the proof by showing that, since $\triangle A'XY \equiv \triangle B'XY$, we have $\angle A'XY \equiv \angle B'XY$, and since $\angle A'XY \equiv \angle B'XY$, we have $\triangle A'XE \equiv \triangle B'XE$. Thus $\angle A'EX$ and $\angle B'EX$ are congruent angles in a linear pair, and therefore are right angles. It follows that $\overleftrightarrow{XY} \perp \overleftrightarrow{AB}$.

• *Problem 5.* A ray $\overrightarrow{BX}$ between $\overrightarrow{BA}$ and $\overrightarrow{BC}$ is the *bisector* of $\angle ABC$ if $\angle ABX \equiv \angle CBX$. Given $\angle ABC$, construct X so that $\overrightarrow{BX}$ is the bisector of $\angle ABC$ (see Fig. 3–18).

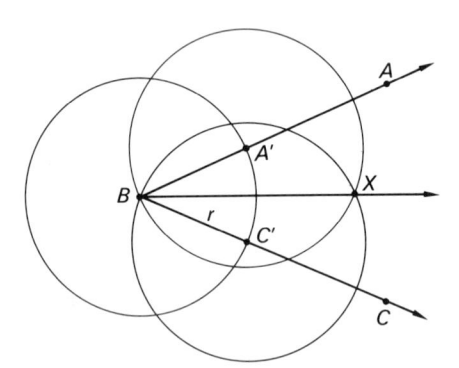

Figure 3–18

Construction 5. Set the compass with any radius r and draw a circle with center at B (Fig. 3–18). This circle intersects $\overrightarrow{BA}$ at a point A' and intersects $\overrightarrow{BC}$ at a point C', by Postulate 4(a). With the same radius r, draw two circles, one with center at A' and the other with center at C'. These circles intersect at B and at a point X (by Postulate 4b) which can be shown is inside $\angle ABC$. The ray $\overrightarrow{BX}$ is the bisector of $\angle ABC$.

Proof of Construction 5. The reader can fill in the proof by assuming that X is inside $\angle ABC$ and then showing that $\triangle A'BX \equiv \triangle C'BX$; therefore $\angle A'BX \equiv \angle C'BX$.

exercise set 3–1

1. In each of the parts of Fig. 3–19, describe $C \cap k$, the intersection of circle C and line k.

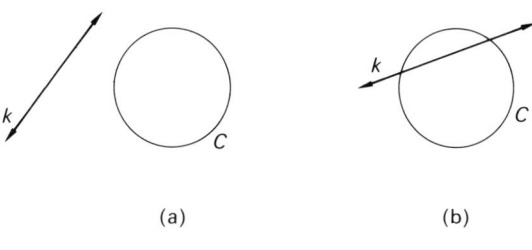

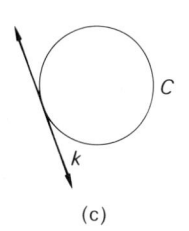

Figure 3–19   (a)   (b)   (c)

2. Draw a line k and a triangle T such that $k \cap T$ is a set consisting of the following. (a) No points (the empty set); (b) 1 point; (c) 2 points; (d) a line segment.

3. Draw a circle C and a triangle T such that $C \cap T$ is a set consisting of the following. (a) No points; (b) 1 point; (c) 2 points; (d) 3 points; (e) 4 points; (f) 5 points; (g) 6 points.

4. In Fig. 3–20 determine
   a) $\overrightarrow{UY} \cap \overrightarrow{YU}$ b) $\overrightarrow{UY} \cup \overrightarrow{YU}$ c) $\overrightarrow{UV} \cup \overrightarrow{YV}$ d) $\overrightarrow{UV} \cap \overrightarrow{YV}$
   e) $\overrightarrow{UV} \cap \overrightarrow{YV}$ f) $\overrightarrow{UV} \cap \overrightarrow{YU}$ g) $\overrightarrow{UV} \cup \overrightarrow{YU}$ h) $\overleftrightarrow{XU} \cup \overleftrightarrow{XU}$
   i) $\overleftrightarrow{XU} \cap \overleftrightarrow{XU}$

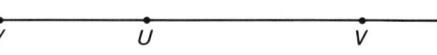

Figure 3–20

5. Is it possible to draw two segments $\overline{AB}$ and $\overline{CD}$ such that $\overline{AB} \cap \overline{CD} = \varnothing$, but $\overleftrightarrow{AB} = \overleftrightarrow{CD}$?

6. Describe the set of points of a triangular region ABC in terms of the intersection of half-planes.

7. Review Chapter 1 by summarizing the definitions of geometric figures given there.

8. Using only Definition 1, Definition 2, Definition 3, the undefined words point, line, plane, and between, and set language, give definitions for these words: (a) triangle; (b) parallel lines in a plane; (c) quadrilateral.

9. In Definition 5 why is it necessary to include the phrase "in a plane"? What figure would be specified by Definition 5 if this phrase were deleted?

10. Two angles in a linear pair are supplementary. If two angles are supplementary, do they form a linear pair? (Refer to Definition 6.)

11. Draw circles C_1 and C_2 with centers at points A and B and with radii of 2 inches and 3 inches, respectively, so that each of the following is true.
    a) A and B are 4 inches apart.
    b) A and B are 6 inches apart.

12. a) In which case—11(a) or 11(b)—did circles C_1 and C_2 intersect at two points? Relate these results to Postulate 4, where a and b are the radii of circles C_1 and C_2 and c is the distance between the centers.
    b) Discuss this case: $a = 4$ inches, $b = 7$ inches, $c = 2$ inches.
    c) Discuss this case: $a = 3$ inches, $b = 3$ inches, $c = 1$ inch.

13. Show that Postulate 2(c) is equivalent to the *triangle inequality:* Each side of a triangle has length less than the sum of the lengths of the other two sides.

14. Prove: For each pair of points A, B, there is exactly one line segment with A and B as its endpoints. [*Hint:* Suppose that there were two segments with A and B as endpoints of each. How would that contradict Definition 2 and Postulate 1(b)?]

15. Prove: If a line intersects a plane but is not contained in the plane, then the line and the plane intersect in exactly one point. [*Hint:* Suppose that line l intersected plane $\mathbb{P}$ in two points R and S. How would the consequence of Postulate 1(e) then contradict the hypothesis that l is not contained in $\mathbb{P}$?]

16. Prove: There is exactly one plane containing a given line and a given point not on the line.

17. Prove: If two lines intersect, then both lines are contained in exactly one plane. [*Hint:* What are reasons for these steps in a proof? Lines l and k intersect in a point Q; k contains a point R different from Q; there is a plane containing l and R; this plane contains l and k; but no other plane contains l and k.]

18. Imagine a space consisting of exactly four points (Fig. 3–21), in which each line consists of exactly two points and each plane consists of exactly three points.
    a) Verify that all six parts of the incidence postulate are satisfied by this space.
    b) How many lines and planes are there in this space?
    c) Verify that the statements of Exercises 15–17 hold true in this space.

3-4　　assumptions about congruent triangles　　85

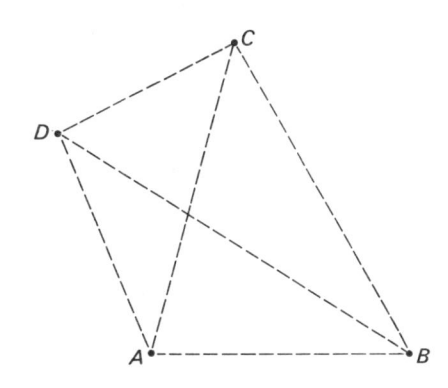

Figure 3–21

19. a) Construct an equilateral triangle PQR whose sides are all congruent to $\overline{AB}$ (Fig. 3–22).
    b) Prove that your construction always yields an equilateral triangle.

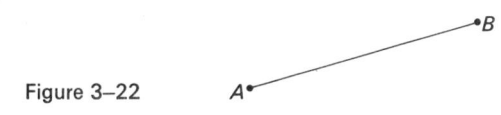

Figure 3–22

20. a) Construct $\triangle PQR$ so that $\overline{PQ} \equiv \overline{AB}$, $\angle QPR \equiv \angle BAC$ and $\angle PQR \equiv \angle ABC$ (see Fig. 3–23).
    b) Is $\angle PRQ \equiv \angle ACB$? Why?

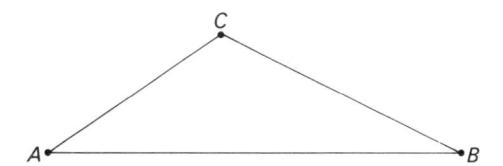

Figure 3–23

21. Describe the following construction and prove the construction: Given any $\triangle ABC$ and $\overrightarrow{XY}$ in a plane, construct points D and E in the plane such that D belongs to $\overrightarrow{XY}$ and $\triangle XDE \equiv \triangle ABC$.

22. A point X which belongs to $\overline{AB}$ is called the *midpoint* of $\overline{AB}$ if $\overline{AX} \equiv \overline{BX}$. Given $\overline{AB}$, construct the midpoint of $\overline{AB}$. Prove your construction.

23. Prove: In $\triangle ABC$, if $\overline{AC} \equiv \overline{BC}$ and $\overrightarrow{CP}$ bisects $\angle ACB$, then $\overleftrightarrow{CP}$ is perpendicular to $\overleftrightarrow{AB}$. (In your proof assume that $\overrightarrow{CP}$ intersects $\overline{AB}$ in a point between A and B.)

24. Complete the proof of Construction 5, page 82. (In your proof assume that X is inside $\angle ABC$.)

25. Prove that the two angles opposite the congruent sides of an isosceles triangle are congruent.

26. a) Given $\overline{AB}$. Construct right triangle PQR so that $\angle PQR$ is the right angle and $\overline{QP} \equiv \overline{QR} \equiv \overline{AB}$ (see Fig. 3–24).
    *b) Construct right triangle STU so that $\angle STU$ is the right angle, $\overline{ST} \equiv \overline{UT}$, and $\overline{SU} \equiv \overline{AB}$.

Figure 3–24

27. Prove that if two angles of a triangle are congruent, the triangle is isosceles.

28. Definition: Two angles, $\angle APB$ and $\angle CPD$, are called *vertical* angles if $\overrightarrow{PA} \cup \overrightarrow{PC} = \overleftrightarrow{AC}$ and $\overrightarrow{PB} \cup \overrightarrow{PD} = \overleftrightarrow{BD}$ or if $\overrightarrow{PA} \cup \overrightarrow{PD} = \overleftrightarrow{AD}$ and $\overrightarrow{PB} \cup \overrightarrow{PC} = \overleftrightarrow{BC}$ (see Fig. 3–25).

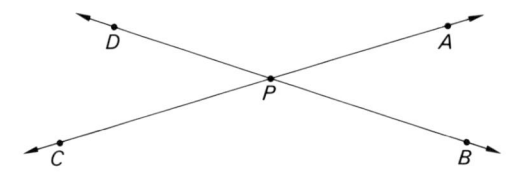

Figure 3–25

Prove: Two vertical angles are congruent. [*Hint:* Show that each of the two vertical angles has the same partner in a linear pair, and recall that the sum of the degree measures of two angles in a linear pair is 180.]

29. Prove: If two line segments $\overline{AB}$ and $\overline{CD}$ bisect each other at point E, then $\overline{AC} \equiv \overline{BD}$.

30. a) Show that in Fig. 3–26 $m(\angle ACF)$ is greater than $m(\angle EBA)$, where $AE = EP$, E is between A and P, and E is the midpoint of $\overline{BC}$. [*Hint:* Since the vertical angles at E are congruent, show that $\triangle ABE \equiv \triangle PCE$; then $\angle EBA \equiv \angle ECP$ and $m(\angle BCD) = m(\angle ECP) + m(\angle PCD)$. Vertical angles at C are congruent, which gives us $m(\angle ACF) = m(\angle ECP) + m(\angle PCD) = m(\angle EBA) + m(\angle PCD)$. The conclusion then follows.]

---

*Starred problems are those which are unusually difficult.

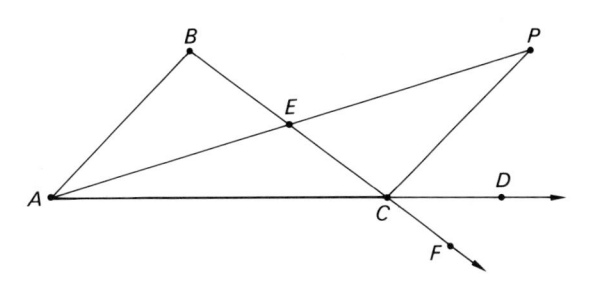

Figure 3-26

†b) By noting that $\angle ACF \equiv \angle BCD$ (why?), restate the result of part (a) as $m(\angle BCD) > m(\angle CBA)$; that is, an exterior angle ($\angle BCD$) of a triangle ($\triangle ABC$) has greater measure than either of its remote interior angles ($\angle CBA$ or $\angle BAC$).

31. Draw an angle POQ (measuring about 60°) and a ray $\overrightarrow{AB}$. Draw a circle C_1 with center O and radius r. Let X and Y be points of intersection of the circle and $\angle POQ$. Draw a circle C_2 of the same radius r and with center at A, intersecting $\overrightarrow{AB}$ at D. Draw a circle C_3 of radius $\overline{XY}$ with center at D and let the intersections of circles C_2 and C_3 be E and F.
    a) Give an argument that circles C_2 and C_3 must intersect in two points.
    b) Find all segments in the figure that are congruent to $\overline{DE}$.
    c) Why is $\triangle XOY \equiv \triangle EAD$?
    d) Find all triangles in the figure that are congruent to $\triangle EAD$.
    e) Find all angles in the figure that are congruent to $\angle XOY$.

32. Construct a 90° angle, a 45° angle, a $22\frac{1}{2}$° angle.

33. Draw a circle with any radius, with center O, and let $\overline{AB}$ be a diameter. Let $\triangle AOD$ and $\triangle BOC$ be two congruent triangles such that $\angle COB$ is a 40° angle and D, C are on the circle. Let E be a point on the circle such that $\angle BOE$ is an 80° angle. Is $\triangle AOE \equiv \triangle DOC$? Why or why not? Can C, D, and E be located on the circle so that $\triangle AOE$ is not congruent to $\triangle DOC$?

teaching questions and projects 3–1

1. Compile a list of objects that children would find useful in a classroom to learn about points, lines, planes, line segments, angles, common plane figures, and common space figures.

---

†Necessary for the proof of Theorem 2 in the next section.

2. Describe classroom activities that help children learn:
   a) to construct congruent line segments and congruent circles using compasses made of strings and pencils;
   b) to construct congruent angles using half-discs.

3. Write a script of a hypothetical classroom discussion in which children move from intuitive ideas of line segments, rays, and angles as idealizations of physical objects to definitions of these figures as sets of points.

4. Make a list of familiar objects that could be found in and around most classrooms that suggest curves, closed curves, and simple closed curves.

5. Review some of the generalizations from experiences with physical objects discussed in Chapter 1. Describe an activity in which children begin to organize and systematize these generalizations in terms of those which depend on others, those which are basic postulates, and so on.

6. At what level or with what background experience as prerequisite should children begin formalizing their geometric intuitions? What factors are relevant to deciding where and when the treatment of postulates and proofs enters the curriculum? Discuss in detail.

7. Describe activities that will help children develop the concept of a line segment as an infinite set of points using the intuitive idea of betweenness.

8. Describe a classroom activity using paper folding that helps pupils:
   a) show that a linear pair of angles that are congruent are also right angles;
   b) show that "square corners" may be used to help draw perpendicular lines.

9. Explain how children may verify, by means of cutouts, that Case AAA (angle-angle-angle) does *not* establish congruence of triangles.

10. Describe a classroom activity that would help children learn to construct perpendicular lines.

11. Describe a classroom activity to help children discover the SSS postulate.

3-5
parallel lines

Let us define parallel lines.

▶ *Definition 10.* Two lines in a plane are parallel if and only if their intersection is the empty set. (We let ∥ denote "is parallel to"; for example, $\overleftrightarrow{AB} \parallel \overleftrightarrow{MP}$.)

We need a test for deciding whether two lines are parallel, since it is not possible to follow the lines forever to check whether they meet. Experiences with drawings suggest that two lines are parallel if there is a line perpendicular to (⊥) both. We state this result as a theorem here, but will give the proof later.

▶ *Theorem 2.* Two lines in a plane are parallel if and only if there is a line in the plane perpendicular to both lines.

It is instructive to see what happens if the second occurrence of the phrase "in the plane" is deleted from the statement of Theorem 2. It then reads:

Two lines in a plane are ∥ if and only if there is a line ⊥ to both.

Does this statement fit our intuition? Imagine the two lines intersecting in a plane, and imagine a line perpendicular to both of them but not in the plane of the two intersecting lines, as shown in Fig. 3–27. Thus the second occurrence of the phrase "in the plane" is crucial in the statement of Theorem 2.

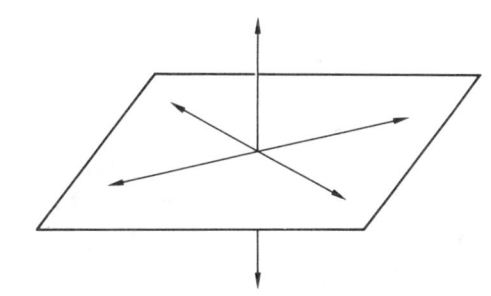

Figure 3–27

In order to prove our criterion for parallelism (Theorem 2), we need an agreement limiting the number of lines that are parallel to a given line through a given point not on the line. Our next postulate, the famous *parallel postulate*, will state that there is exactly one such line.

Postulate 6. (Parallel postulate) Through a given point P not on a line $\overleftrightarrow{AB}$ there is exactly one line parallel to $\overleftrightarrow{AB}$.

As soon as Theorem 2 is proved we shall be able to construct lines parallel to given lines as follows:

* *Problem 6.* Given $\overleftrightarrow{AB}$ and P not belonging to $\overleftrightarrow{AB}$, construct $\overleftrightarrow{PQ}$ so that $\overleftrightarrow{PQ} \parallel \overleftrightarrow{AB}$.

Construction 6. Construct $\overleftrightarrow{PR}$ so that $\overleftrightarrow{PR} \perp \overleftrightarrow{AB}$, by Construction 4. Then construct $\overleftrightarrow{PQ}$ so that $\overleftrightarrow{PQ} \perp \overleftrightarrow{PR}$, by Construction 3. Then $\overleftrightarrow{PQ} \parallel \overleftrightarrow{AB}$.

Proof of Construction 6. Since $\overleftrightarrow{PQ} \perp \overleftrightarrow{PR}$ and $\overleftrightarrow{PR} \perp \overleftrightarrow{AB}$, we conclude that $\overleftrightarrow{PQ} \parallel \overleftrightarrow{AB}$, by Theorem 2.

Proof of Theorem 2. We need to prove two separate parts of the theorem:

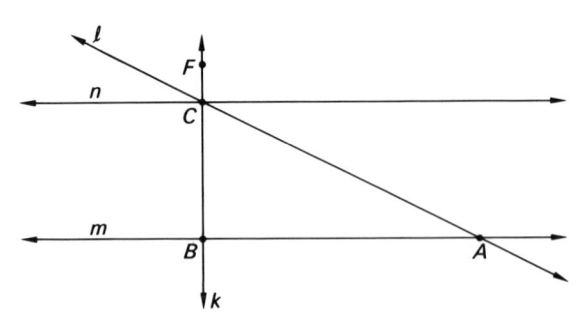

Figure 3–28

1) Given that m and n are parallel lines (see Fig. 3–28), then there is a line k in the plane of m and n such that $k \perp m$ and $k \perp n$.
2) Given that m, n, k are lines in a plane such that $k \perp m$ and $k \perp n$, then $m \parallel n$.

Proof of (1). Let $m \parallel n$. Choose any point B on m and construct a line k that contains B and is perpendicular to m, by Problem 3.

Then there are two cases to consider, one of which must be true:
a) k does not intersect n, or
b) k does intersect n.

In Case (a), by Definition 10, $k \parallel n$; we then have distinct lines k and m, both through B and both parallel to n. This is impossible, according to Postulate 6. Thus Case (b) must hold, and lines k and n intersect in some point C.

Again there are two cases to consider, one of which must be true:
c) k is not perpendicular to n, or
d) k is perpendicular to n.

In Case (c), we have $m \parallel n$, $k \perp m$, k intersects n in C, and since n is not perpendicular to k, we can construct line l through C that is perpendicular to k. Then, by Postulate 6, l cannot be parallel to m (since we already have one line n through C parallel to m), and as a consequence l intersects m in some point A.

As matters now stand, we have two distinct lines m and l, each through A and each perpendicular to k. Let us show that this is not possible. We do this by appealing to the result of Exercise 30 on page 86. Assume [as Case (c) leads us to assume] that $\angle ACF$ and $\angle CBA$ are both right angles and therefore congruent. But in $\triangle ABC$, Exercise 30 shows that the exterior angle, $\angle ACF$, has greater measure than remote interior angle $\angle CBA$. Thus these two angles cannot be congruent; only one of these angles can be a right angle; and Case (c) leads to an impossible situation. We have shown that through C only one line is perpendicular to line k; thus we must accept the other alternative, Case (d). This proves part (1) of the theorem.

Proof of (2). Let $k \perp m$ and $k \perp n$. Let k intersect m at B and intersect n at C. Again we see that one of two cases must hold: either m is not parallel to n, or $m \parallel n$. If m is not parallel to n, then m intersects n in some point T, and we again have the situation of two distinct lines through a point, each perpendicular to the same line. We showed in part (1) of the proof that this is impossible. Hence $m \parallel n$, and the theorem is proved.

Using Theorem 2 it is now possible to prove a theorem that was stated and used in Section 3–4 without proof (see Exercise 2 in Exercise Set 3–2).

▶ *Theorem 3.* The sum of the degree measures of the angles of a triangle is 180.

Recall that we anticipated this theorem in order to prove the AAS case of congruent triangles (Theorem 1), so that we could prove some geometric constructions. When the reader proves the angle-sum theorem, he should check his steps to be sure that he does not use any previous results, such as AAS congruence of triangles, that depend on the angle-sum theorem. Note that Theorem 2, the criterion for parallelism, is a key to the proof of the angle-sum theorem; did the proof of Theorem 2 use the angle-sum result? If it had, we would have been guilty of circular reasoning.

In the proof of Theorem 2 we were able to show that in a plane there is only one line through a given point that is perpendicular to a given line. One might question why we needed to assume a corresponding result for parallel lines (as we did in Postulate 6). Why not *prove* that through a given point only one line can be drawn parallel to a given line? Surprisingly, this cannot be proved, and we are free to assume as many parallels to a line through a point as we wish. When we take Postulate 6 as our assumption, the resulting geometry is called Euclidean. When we assume no parallels or many parallels can be drawn through a point parallel to a given line, the resulting geometries are called non-Euclidean. It was shown by Lobachevski (1793–1856), Riemann (1826–1866), and Bolyai (1802–1860) that such non-Euclidean geometries are just as consistent as Euclidean geometry, and since then these geometries have been found useful as mathematical models in theoretical physics.

3–6
looking more closely at proofs

In Sections 3 to 5 we listed some statements that we could assume as postulates in order to prove some of the theorems of plane geometry. Our objective was not to give a rigorous development of all of plane geometry; this is not a viable objective in the elementary program. Our objective was rather to illustrate the deductive process in short chains of reasoning. In other words, we were illustrating how to form proofs.

When we begin to ask, "How do I know it is true in all cases?" we are beginning to understand the need for proofs. When we ask, "If I accept those statements, am I forced to accept this statement?" we are beginning to understand how a proof is made.

If the form of a set A of statements is such that it leads us to another statement B and if we have accepted the statements A, then we must accept statement B as well. Thus, in order to prove a statement, we must decide beforehand:

1. what statements are accepted without proof (postulates),
2. what statements are accepted because they were proved as consequences of the postulates (theorems), and
3. what the terms in the statements mean (definitions).

The statement to be proved and each of the statements in the postulates and the theorems can be put in a form called a *conditional statement* or *implication:*

$$\text{If \_\_\_\_\_, then \_\_\_\_\_.}$$

The "if" part of the statement is the *hypothesis* or *antecedent*. The "then" part is the *conclusion* or *consequence*. For example, consider the statement:

▲ If (k, m, n are lines in a plane and $m \parallel n$ and $k \perp m$), then ($k \perp n$).

We write the hypothesis and conclusion inside parentheses to emphasize them.

Definition 10, for example, says in part that:

If (the intersection of two lines in a plane is the empty set), then (the lines are parallel).

Definition 10 also says that:

If (two lines in a plane are parallel), then (their intersection is the empty set).

Note that statements of definitions may always be stated in the form

(A) *if and only if* (B).

This refers to two statements: "if (A), then (B)," and "if (B), then (A)."

Let us write Statement ▲ above in the form: If A, then C. To prove Statement ▲, we would first prove two other statements:

1. If $m \parallel n$ and $k \perp m$, then $m \parallel n$ and $k \perp m$ and k intersects n. (If A, then B.)
2. If $m \parallel n$ and $k \perp m$ and k intersects n, then $k \perp n$. (If B, then C.)

If we can prove statements (1) and (2), then we agree that the statement "If A, then C" has been proved:

If $m \parallel n$ and $k \perp m$, then $k \perp n$. (If A, then C.)

This is statement ▲.

In general, we agree that if A, B, C are any statements, and if we accept the statements "If A, then B" and "If B, then C," then we must also accept the statement "If A, then C." This agreement is called the Transitivity of Implication. We call this a direct proof of the statement "If A, then C." Any chain of implications of the form

if A, then A_1; if A_1, then A_2; if A_2, then A_3; ... ; if A_k, then C,

in which each implication is accepted as a postulate, a theorem or a definition, leads to our acceptance of

if A, then C,

and constitutes a *direct proof* of "if A, then C."

There are more forms of statements that lead automatically to others, and some that do not. For example, if we accept the statement

i) if $m \cap n \neq \emptyset$, then $m \not\parallel n$ (m and n lines in a plane),

we are then forced to accept the statement

ii) if $m \parallel n$, then $m \cap n = \emptyset$ (m and n lines in a plane).

We say that implications (i) and (ii) are *contrapositives* of each other. The acceptance of any implication "if A, then C" forces us to accept its contrapositive "if not C, then not A," and vice versa, regardless of the content of the implication, but strictly due to the form of a contrapositive.

On the other hand, even though we accept the statement

iii) if $m \parallel n$, then $m \cap n = \emptyset$ (m and n lines in space),

we are not forced to accept its converse statement

iv) if $m \cap n = \emptyset$, then $m \parallel n$ (m and n lines in space),

since we can find lines m and n in space that are not parallel, yet do have an empty intersection. We say that (iii) and (iv) are *converse* statements, and the acceptance of an implication "if A, then C" *does not* force us to accept its converse, "if C, then A." Occasionally we can prove as theorems both an implication and its converse, as in Theorem 2. In that case we have proved "A if and only if C." But our having proved "if A, then C" tells us nothing about "if C, then A." The latter must be proved or disproved separately.

Theorem 2 is particularly interesting because it illustrates an *indirect* proof. In the proof of Theorem 2 we used an argument in which an assumption that one of two possible alternatives is true leads us to an impossible statement, thereby forcing us to accept the other alternative as true.

Such indirect arguments are used freely in everyday affairs. For example, a girl would like to prove to her satisfaction that her father is at home. She knows that either he is at home or he has gone to work. She assumes that he has gone to work; then she looks in the garage and sees his car there. But it is impossible that he could have gone to work without his car. Thus the assumption that he has gone to work leads to an impossible situation, and she must conclude that he is still at home.

In some cases an indirect argument is the only one that will yield a proof. Consider the statement:

If a is an integer and a^2 is even, then a is even.

A direct proof would require testing every even square, which clearly cannot be done. In an indirect proof the assumption that a is odd will lead to a contradiction of the given fact that a^2 is even (since the square of an odd integer is odd). This forces us to conclude that a is *not* odd; hence it must be even.

As another example, consider the statement (Exercise 15 in Exercise Set 3–1):

If a line l intersects plane $\mathbb{P}$ but l is not contained in $\mathbb{P}$, then l intersects $\mathbb{P}$ in exactly one point.

An indirect proof begins with the statement that since l intersects $\mathbb{P}$, the intersection consists either of exactly one point or of more than one point, but not both. Assume that the intersection contains at least two points Q and R. Then two points of l are in $\mathbb{P}$ and by Postulate 1(e) the whole line $\overleftrightarrow{PQ}$ is contained in $\mathbb{P}$. This contradicts the hypothesis that l is not contained in $\mathbb{P}$, and our assumption must be rejected. Our only other alternative—that there is exactly one point of intersection—must be accepted.

To summarize, an indirect proof of the implication "if A, then C" is an argument showing that the acceptance of the assumption "A and not C" leads to a contradiction of some accepted statement.

3–7
right triangles

If one angle of a triangle has degree measure 90, the triangle is called a right-angled triangle, or *right triangle*. The sides of the triangle contained in the sides of the right angle are called the *legs*, and the third side (opposite the right angle) is called the *hypotenuse*. We shall conclude this chapter by proving two important theorems about right triangles and then using the theorems to construct segments whose measures are irrational numbers.

▶ *Theorem 4.* If A, B, C are points on a circle (Fig. 3–29) and $\overline{AB}$ is a diameter of the circle, then $m(\angle ACB) = 90$.

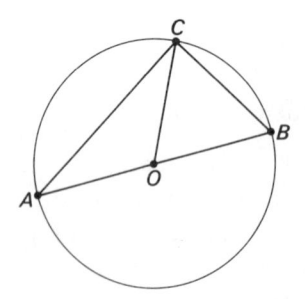

Figure 3–29

Proof of Theorem 4. Since the center of the circle is on $\overline{AB}$, then $\overline{OA}$, $\overline{OB}$, and $\overline{OC}$ are all congruent. Thus $\triangle OAC$ and $\triangle COB$ are isosceles triangles, and therefore $\angle CAO \equiv \angle ACO$ and $\angle OCB \equiv \angle OBC$. Then:

$m(\angle ACB) = m(\angle ACO) + m(\angle OCB)$ because of Postulate 3(b) and the fact that $\overrightarrow{CO}$ is between $\overrightarrow{CA}$ and $\overrightarrow{CB}$.

$m(\angle ACB) = m(\angle CAO) + m(\angle OBC)$ because by definition congruent angles have equal measures.

$m(\angle ACB) = 180 - m(\angle ACB)$ because the sum of the degree measures of the angles of a triangle is 180, $m(\angle ACB) + m(\angle ACB) = 180$, and $m(\angle ACB) = 90$.

The theorem is proved. (Note that the last three steps of the proof rely on properties of operations on real numbers.)

▶ *Theorem 5.* (Pythagorean Theorem) Given that $\angle ABC$ is a right angle in $\triangle ABC$, then $(AC)^2 + (BC)^2 = (AB)^2$.

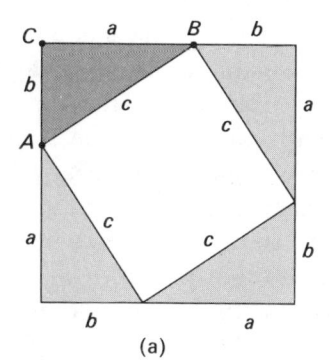

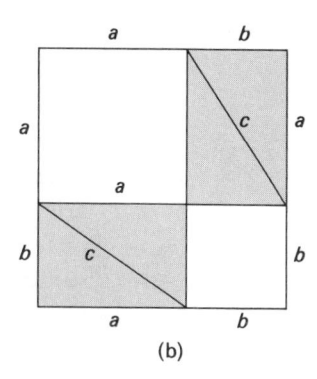

Figure 3–30

(a)   (b)

Proof of Theorem 5. Let $AC = b$, $BC = a$, and $AB = c$. Draw two congruent squares (Fig. 3–30), each having sides of length $a + b$, and partition the squares as shown in the figure.

The eight shaded triangles in the two squares are all congruent to each other (why?), and hence have the same area.

Also, since each of the eight shaded triangles is congruent to $\triangle ABC$ (why?), each has hypotenuse of length c.

We shall prove that the unshaded quadrilateral in part (a) is a square: Since its sides all have length c, we need only show that all its angles are right angles. We know that each angle in the quadrilateral has measure of

$$180 - [m(\angle CBA) + m(\angle CAB)]. \quad \text{(Why?)}$$

But

$$m(\angle CBA) + m(\angle CAB) = 180 - m(ACB) = 180 - 90 = 90.$$

Hence each angle in the quadrilateral has measure 90 and the quadrilateral is a square with area c^2.

In part (b), the area of the shaded regions is the same as the area of the shaded region in part (a). (Why?) Hence the areas of the unshaded regions of the two figures are the same. (Why?) The area of the unshaded region in part (a) is c^2; in part (b), the area of the unshaded regions is $a^2 + b^2$. Therefore $c^2 = a^2 + b^2$.

The theorem is proved.

remarks concerning theorems 4 and 5 If these theorems are proved in an elementary school program (this depends on the level of maturity attained by the students), the proofs will be culminations of much exploratory work and much free experimentation with right triangles. After children draw many triangles in a circle, with one side of the triangle a diameter, and after they measure the angles of the triangles, the conjecture will arise that all such triangles are right triangles. Only then would an attempt to prove this conjecture be made, and only if the students need a proof to satisfy them.

The same is true of the proof of the Pythagorean theorem. Many special cases of right triangles are first encountered—such as the 3, 4, 5 triangle, the 5, 12, 13 triangle, and the 7, 24, 25 triangle, before pupils suspect that a relation among the squares of the lengths of the sides is generally true. Here there are a great number of opportunities for discovery, whether or not the activity terminates in a formal proof.

constructing a segment with irrational measure

• *Problem 7.* Let $AC = 1$, $CB = 1$, and construct a right triangle $\triangle ABC$ with $m(\angle C) = 90$ (see Fig. 3–31). Find AB.

Construction 7. Set a compass with radius $r = 1$, in any unit of length. On a line $\overleftrightarrow{XY}$ construct a segment $\overline{AC}$ of length 1. At

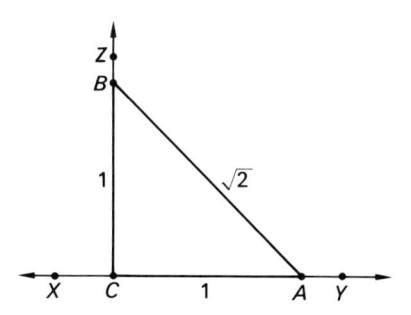

Figure 3-31

C erect a perpendicular $\overrightarrow{CZ}$ to $\overline{AC}$. On $\overrightarrow{CZ}$ construct a point B such that $CB = 1$. Then $m(\angle ACB) = 90$, and $\triangle ABC$ is the required triangle. By the Pythagorean theorem,

$$(AB)^2 = (AC)^2 + (CB)^2 = 1 + 1 = 2.$$

Hence $\overline{AB}$ is a segment such that the square of its measure is 2. In other words, $AB = \sqrt{2}$.

We promised in Chapter 2 that we would construct a segment whose measure is an irrational number. We claim that the segment $\overline{AB}$ constructed in Problem 7 is such an example.

Again this topic provides opportunity for experimentation and discovery. The search for a rational number whose square is 2 can lead to the idea of nested intervals, as described in Chapter 2. For example, we see that $\sqrt{2}$ cannot be a whole number, since $1^2 = 1$ and $2^2 = 4$, whereas 2 is between 1 and 4. So we try to find a rational number whose square is 2. We find that

$(1.4)^2 = 1.96$ and $(1.5)^2 = 2.25,$ so that $1.4 < \sqrt{2} < 1.5,$

$(1.41)^2 = 1.9881$ and $(1.42)^2 = 2.0164,$
so that $1.41 < \sqrt{2} < 1.42$, etc.

At each step we find two rational numbers that are closer to $\sqrt{2}$ and straddling $\sqrt{2}$, but we never find a rational number whose square *is* 2. We suspect that there is no rational number whose square is 2.

▶ *Theorem 6.* There is no rational number whose square is 2.

Proof of Theorem 6. We shall use an indirect argument.

If we assume that there is a rational number, say $\frac{a}{b}$, such that $\left(\frac{a}{b}\right)^2 = 2$, then $a^2 = 2b^2$, where a and b are some positive integers. It is known from arithmetic that each positive integer can be expressed in exactly one way as a product of prime numbers. Thus a^2 is a product of an even number of primes, and b^2 is the product of an even number of primes. (Why?) But 2 is also a prime number, so that $2b^2$ can be expressed in exactly one way as a product of an *odd* number of primes.

This leads us to the situation of having a number (denoted by both a^2 and $2b^2$, since $a^2 = 2b^2$) expressed both as a product of an even number of primes and of an odd number of primes. But there is exactly one prime factorization of each positive integer. Since our assumption leads us to a contradiction, we must discard the assumption and conclude that there is no rational number whose square is 2.

The theorem is proved.

exercise set 3–2

1. *Definition:* A line in a plane that intersects each of two given lines in the plane in exactly one point is called a *transversal* of the two lines (Fig. 3–32).

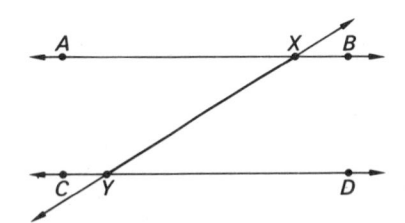

Figure 3–32

a) Prove: If two lines $\overleftrightarrow{AB}$ and $\overleftrightarrow{CD}$ are cut by a transversal at X and Y, respectively, and if $\angle AXY \equiv \angle DYX$, then $\overleftrightarrow{AB} \parallel \overleftrightarrow{CD}$. $\angle AXY$ and $\angle DYX$ are called *alternate interior angles*. [*Hint:* Assume $\overleftrightarrow{AB}$ and $\overleftrightarrow{CD}$ are not parallel, and thus meet in a point T. Then in $\triangle XTY$ the exterior angle at Y is congruent to the remote interior angle at X. How does this contradict the result of Exercise 30 on page 86?]

b) Prove: If two parallel lines $\overleftrightarrow{AB}$ and $\overleftrightarrow{CD}$ are cut by a transversal at X and Y, respectively, then $\angle AXY \equiv \angle DYX$. [*Hint:* Given

3–7   right triangles   101

∠DYX, there is exactly one line k through X for which the alternate interior angles are congruent. Then by part (a), k ∥ $\overleftrightarrow{CD}$. But what does Postulate 6 say about the number of lines through X that are parallel to $\overleftrightarrow{CD}$?]

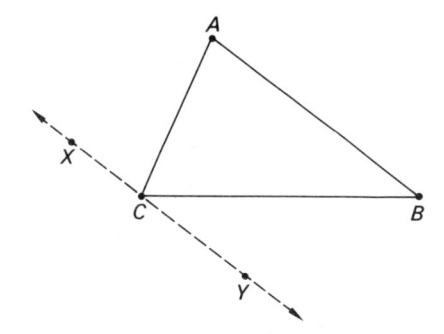

Figure 3–33

2. Use the result of Exercise 1 to prove that the sum of the degree measures of the angles of a triangle △ABC is 180. [Hint: Construct a line $\overleftrightarrow{XY}$ through C and parallel to $\overleftrightarrow{AB}$, as shown in Fig. 3–33. Apply Exercise 1 to show that

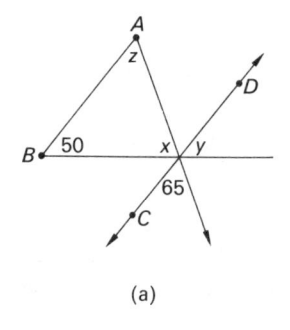

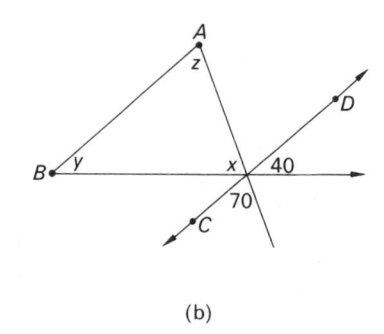

∠ABC ≡ ∠YCB   and   ∠BAC ≡ ∠XCA.

Also ∠BCA ≡ ∠BCA. What is m(∠YCB) + m(∠BCA) + m(∠ACX)?]

Figure 3–34

(a)   (b)

3. In Fig. 3–34, given that $\overleftrightarrow{AB}$ is parallel to $\overleftrightarrow{DC}$, find the measures, in degrees, of ∠x, ∠y, and ∠z.

4. a) Without using a protractor determine the measures, in degrees, of ∠x, ∠y, ∠z, and ∠w in Fig. 3–35.
   b) Which two lines in Fig. 3–35 are parallel? Why?

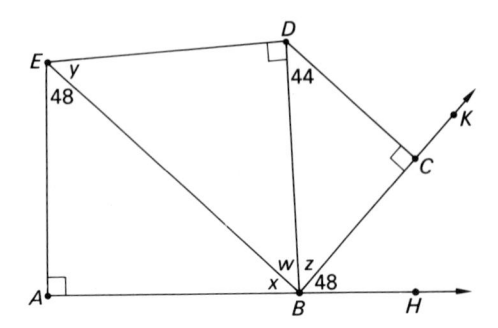

Figure 3–35

c) $\overleftrightarrow{ED}$ and $\overleftrightarrow{AB}$ intersect in a point F not shown in Fig. 3–35. (Why?) Find $m(\angle EFA)$.

d) $\overleftrightarrow{EA}$ and $\overleftrightarrow{BC}$ intersect in a point G not shown in Fig. 3–35. Find $m(\angle AGB)$.

5. Change each of these true statements to the form "if ———, then ———."

   a) Every square is a rectangle.
   b) A rigid motion is a rotation if it leaves exactly one point of the plane fixed.
   c) There are clouds in the sky whenever it is raining.

6. a) For each statement in Exercise 5, now in the form "if ———, then ———," write its converse statement.
   b) Are we forced to accept the converse statements? Which, if any, are true?

7. a) For each statement in Exercise 5, now in the form "if ———, then ———," write its contrapositive statement.
   b) Are we forced to accept the contrapositive statements? Verify that each is true.

8. Prove the following. (Some proofs may require indirect arguments.)
   a) If t is an integer such that t^2 is odd, then t is odd. [Hint: n is even if there is an integer m such that $n = 2m$. Also, r is odd if there is an integer s such that $r = 2s + 1$.]
   b) If t is an odd integer and s is an even integer, then $s + t$ is an odd integer and st is an even integer.
   c) If t and s are integers and ts is odd, then t is odd and s is odd.
   d) If t and s are odd integers, then $t + s$ is even.

9. a) Define these words: rectangle, diagonal of a rectangle, bisecting line segments.
   b) Prove: The diagonals of a rectangle bisect each other.

10. State the converse of Theorem 1, page 79. Is the converse true? Why?

11. State the converse of the theorem in Exercise 27, Exercise Set 3–1. Do you believe the converse is true? If so, prove it.

12. Which of the following triples of numbers can be the measures of sides of a right triangle?
   a) 4, 3, 5
   b) 12, 7, 11
   c) 2, 2, 4
   d) 24, 7, 25
   e) 24, 10, 26
   f) 9, 40, 41

13. Construct a segment whose measure is $\sqrt{3}$. [*Hint:* Construct a right triangle with its sides measuring 1 and $\sqrt{2}$, respectively.]

14. Prove that $\sqrt{3}$ is not a rational number. (Use the proof of Theorem 6 as a guide.)

15. Repeat Exercises 13 and 14 for: (a) $\sqrt{5}$, (b) $\sqrt{7}$.

16. Following the procedure used on page 99 to find rational numbers straddling $\sqrt{2}$, find a rational number in decimal form (with two decimal places) that best approximates: (a) $\sqrt{3}$, (b) $\sqrt{5}$, (c) $\sqrt{7}$.
   d) Explain why there is no end to this procedure for approximating $\sqrt{3}$ with rational numbers.

17. It is known that the real number π is irrational. Its decimal representation does not terminate, and the first 15 places in this decimal are 3.141592653589793 . . . .
   a) Sometimes $\frac{22}{7}$ is used as a rational approximation for π. Change $\frac{22}{7}$ to decimal form and compare it with the decimal representation of π given above. They agree to how many places?
   b) Sometimes 3.14 is used as a rational approximation for π. Which is the better approximation for π: $\frac{22}{7}$ or 3.14?

18. Show that $3\frac{10}{71} < \pi < 3\frac{10}{70}$.

19. Draw the number line and construct points on it whose coordinates are $\sqrt{2}, \sqrt{3}, \sqrt{4}, \sqrt{5}, \sqrt{6}, \sqrt{7}, \sqrt{8}, \sqrt{9}, \sqrt{10}$. [*Hint:* Let the length of the segment from the 0-point to the 1-point be the unit. With this unit construct a right triangle with sides measuring 1, 1, $\sqrt{2}$. Then on this triangle construct another right triangle with sides measuring $\sqrt{2}$, 1, $\sqrt{3}$, etc., as in Fig. 3–36.]

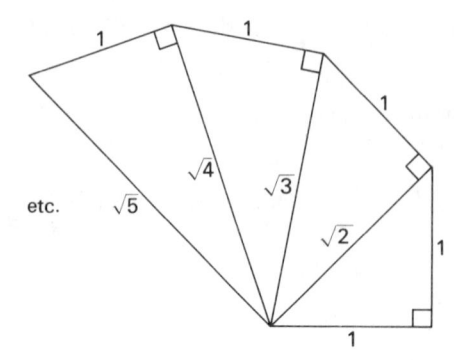

Figure 3–36

20. State the converse of Theorem 5. Do you believe it is true? If so, prove it. [*Hint:* Given $\triangle ABC$ with sides of measures a, b, c such that $a^2 + b^2 = c^2$. Then construct right triangle $A'B'C'$ with $A'C' = b$, $B'C' = a$. Let $d = A'B'$. Why is it true that $a^2 + b^2 = d^2$? Then why is it true that $c^2 = d^2$? Then why does $c = d$ and $\triangle A'B'C' \equiv \triangle ABC$?]

teaching questions and projects 3–2

1. How can a folded square corner be used to help children understand Theorem 2?

2. Describe a classroom activity that would help children learn to construct parallel lines.

3. How can a folded square corner be used to show that if a line is perpendicular to a given line, it is perpendicular to a line parallel to a given line?

4. Gather some simple examples of tile diagrams that suggest the Pythagorean relation.

5. Plan a series of activities in the measurement of sides and angles of triangles that will help children discover the Pythagorean relation.

6. How can you help children understand that there are numbers that are not rational? Include in your discussion the interplay between the problem in geometry of constructing a segment the square of whose measure is 2 or 3 or 5 and the problem in arithmetic of finding a rational number whose square is 2 or 3 or 5.

7. A formal proof has many of the characteristics of a game; we operate by logical rules. We have a starting point, the postulates, and a goal, the theorem. Each deduction is a series of steps, each justified by a rule of the game, from the postulates to the concluding theorem. Construct examples of games in which a word (made up of a string of letters) is the single postulate and the logical rule permits a change of one letter. The theorem is another word. (For example, begin with SEEM; derive BOAT.

   SEEM → TEEM → TEAM → BEAM → BEAT → BOAT.)

chapter four

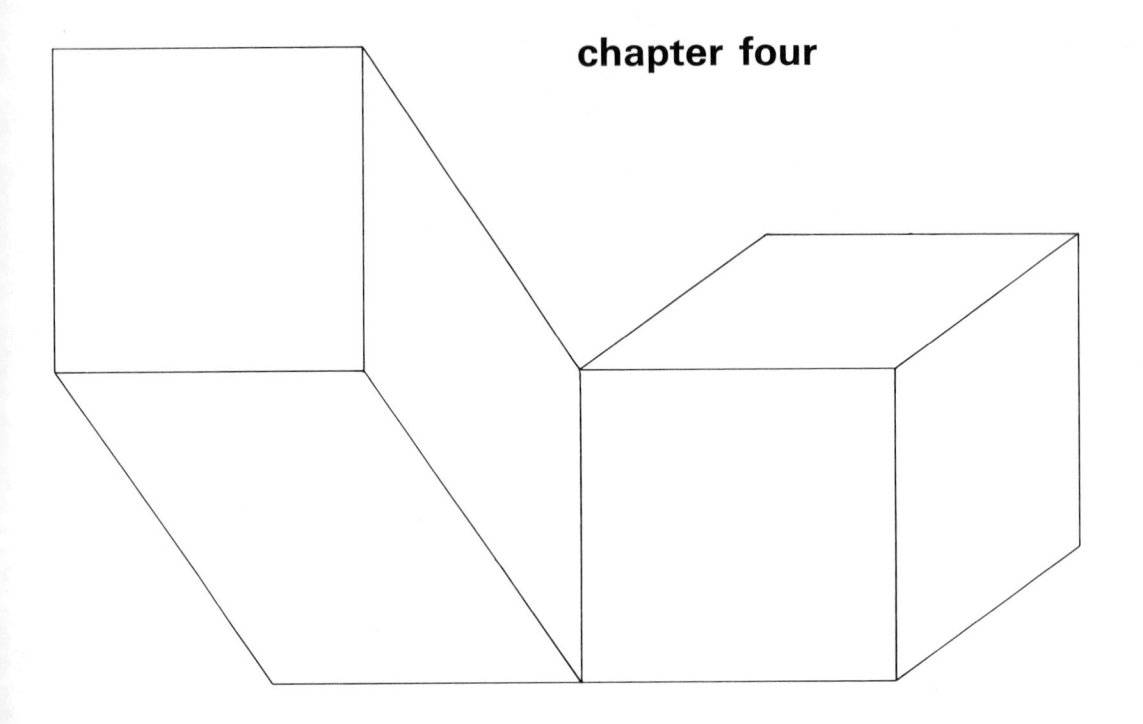

functions in geometry: motions in the plane

4-1
introduction

Throughout the study of geometry, indeed throughout the study of all mathematics, there are certain common concepts that tend to unify the subject. One such unifying thread in mathematics is the concept of function. As a child begins the study of mathematics, he encounters functions in many and varied forms because the concept of functions is not only one of the most basic in mathematics but is also one of the simplest ideas. Although an elementary school teacher may or may not choose to use the word function in his classroom, he should be aware that most of the standard portions of school mathematics are examples of this concept. Thus the notion of a function is central in the program and should be used for unifying many different ideas. This may be done explicitly or implicitly. If it is done implicitly, the particular characteristics of a relationship that make it a function are emphasized even though the students do not explicitly define function.

Mathematics is concerned not only with quantitative descriptions of objects but with the interrelations among sets of objects. It is the interrelations between objects that make such descriptions meaningful. For example, real numbers themselves would be of little interest unless there were operations which relate some numbers to others; geometric figures themselves would not merit serious study unless there were the possibility of establishing interrelations among them.

We shall define a function formally in Section 4–3. For now, we shall consider it a special kind of assignment between two sets. If to each element in one set we assign exactly one element in the other set, we call this assignment a *function*. The elements of the sets may be any objects. For example, in arithmetic we consider sets of numbers and in geometry we study sets of points.

4–2
examples of functions in elementary mathematics

one-to-one matching One of the earliest experiences in learning arithmetic is to pair the elements of two sets of objects to decide whether one set has more elements than the other or whether one set has just as many elements as the other. To each element in the one set, call it P, we assign exactly one element in the other set, say Q (so that no element of Q is assigned to more than one element of P). If this is done and there are elements in Q left over (see Fig. 4–1), then Q has more elements than P.

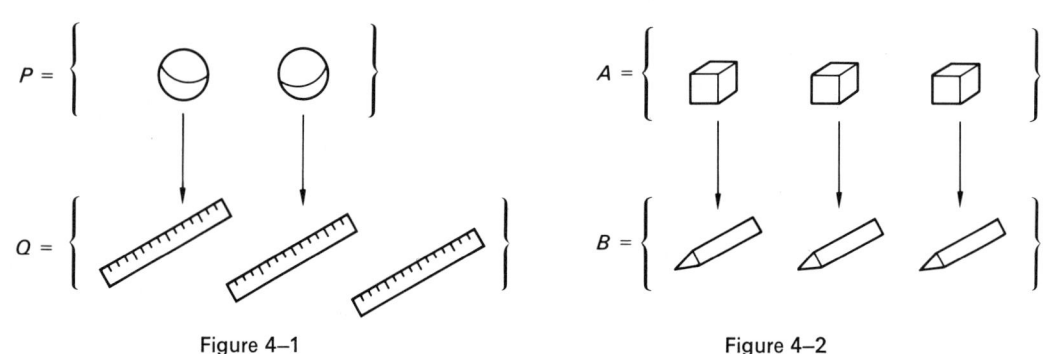

Figure 4–1

Figure 4–2

If, on the other hand, to each element in some set A we can in the same way assign exactly one element in a set B so that no elements of B are left over (see Fig. 4–2), then set A has exactly as many elements as set B. In this latter case, we say that sets A and B are *equivalent*.

Whatever language is used to describe this one-to-one pairing process, it should include the idea that:

The sets A and B are related so that to each element of A there is assigned *exactly one* element of B.

assigning whole numbers to sets To define the set of whole numbers, we collect classes of equivalent sets and make an important abstraction. To each finite set we assign a number which describes the whole class of sets equivalent to this set. To the set $\{X, \theta\}$, for example, we assign the number 2; to the set $\{*\}$ we assign the number 1; to the set $\{a, b\}$ we assign the number 2; to the empty set we assign the number 0; etc. When two sets are equivalent we say that they belong to the same *equivalence class* and we assign the same number to each of the sets in this class. For example, the number 3 is assigned to each set of the class of all sets equivalent to $\{X, \theta, *\}$.

Again, whatever the language used, it will indicate that for the set A of all finite sets and the set B of all whole numbers:

The sets A and B are related so that to *each* element of A (each finite set) there is assigned *exactly one* element of B (a whole number).

Note that the two examples—that of pairing the elements of two sets one-to-one and that of assigning numbers to finite sets—have a common characteristic. To every element in the first set we assign *exactly one* element of the second set. This is required if the assignment is to be a function. But the examples differ in another respect. In pairing elements one-to-one, as in the first example, we do not assign any element of the second set to more than one element of the first set. This is characteristic of one-to-one matching, but not generally required of a function. For we shall see that it is not true of the second example.

In assigning whole numbers to finite sets (in the second example) we assign the same number to many different sets. For example, the set of a pair of shoes, the set of your ears, the set of your eyes, all are sets to which the same number, 2, is assigned. But it *is* the case that the requirement for calling an assignment a function is met; to each finite set we assign one and only one whole number. Thus we see that functions can be one-to-one assignments (Fig. 4–3) or many-to-one assignments (Fig. 4–4).

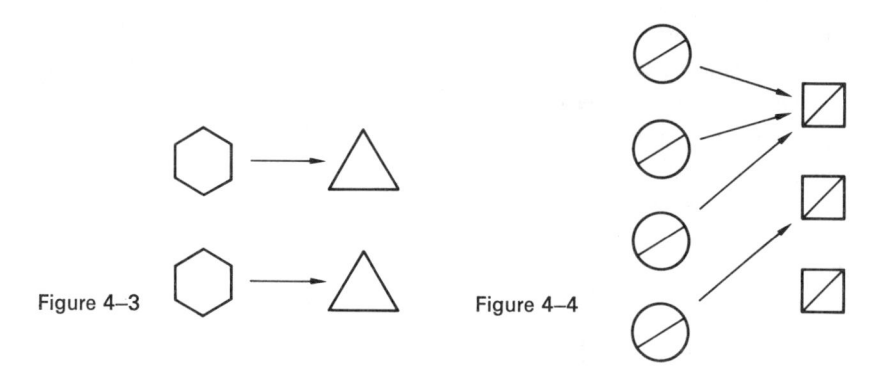

Figure 4–3

Figure 4–4

operations Once we have assigned numbers to finite sets, a next step in arithmetic is to define binary operations on those numbers. Consider the addition of whole numbers as an example. Recall that addition of whole numbers is based on the idea of the union of two disjoint, finite sets. For example,

$$\{a, b\} \cup \{c, d, e\} = \{a, b, c, d, e\},$$

and since we assign the number 2 to the first set and the number 3 to the second set and the number 5 to the union, we conclude that $2 + 3 = 5$. In this example, addition assigns 5 to the ordered pair (2, 3). Addition also makes these assignments:

(2, 7) $\xrightarrow{+}$ 9,   (15, 3) $\xrightarrow{+}$ 18,   (8, 6) $\xrightarrow{+}$ 14,   (5, 0) $\xrightarrow{+}$ 5.

In general, addition assigns a number c to an ordered pair of numbers (a, b). Thus whenever an ordered pair of whole numbers (the *addends*) is given, addition assigns to that pair a whole number (the *sum*). [*Note:* (2, 3) and (3, 2) are the same pair of numbers but they are different *ordered pairs* because they have different first entries or different second entries.]

Is this assignment a function? In this example, set A is a set of ordered pairs of whole numbers. Does addition assign to every ordered pair of whole numbers *exactly one* whole number? The answer is of course yes, and thus addition is another example of a function.

The operation of addition of whole numbers assigns to *each* element of *A* (each ordered pair of whole numbers) *exactly one* element of *B* (a whole number).

Is the operation of addition an example of a one-to-one or a many-to-one function? We note, for example, that the number 5 is assigned to many pairs such as (2, 3), (3, 2), (4, 1), (0, 5). Thus addition is a many-to-one function. But for any pair, there is exactly one sum.

$$(2, 3) \xrightarrow{+} 5 \qquad (6, 4) \xrightarrow{+} 10 \qquad (6, 5) \xrightarrow{+} 11$$
$$(4, 1) \xrightarrow{+} 5 \qquad (2, 8) \xrightarrow{+} 10 \qquad (8, 7) \xrightarrow{+} 15$$
$$(0, 5) \xrightarrow{+} 5 \qquad (8, 2) \xrightarrow{+} 10 \qquad (8, 0) \xrightarrow{+} 8$$

We see that the binary operation of addition of whole numbers is a function. This suggests that we can generally define a binary operation on a set as a function.

A binary operation on a set *S* is a function from $S \times S$ to *S*.

The notation $S \times S$ refers to the cartesian product of set *S* and set *S*.† Thus the first set in the function consists of all possible ordered pairs of the elements in set *S*. To each of these ordered pairs a binary operation assigns exactly one element of set *S*.

---

† The cartesian product of two sets is a set of ordered pairs. Suppose that we have two sets *A* and *B*. The cartesian product $A \times B$ is the set of all ordered pairs (*a*, *b*) such that *a* belongs to *A* and *b* belongs to *B*.

For example, if $A = \{x, y, z\}$ and $B = \{1, 2\}$, then the cartesian product of *A* and *B* is the set of all possible pairings of elements where the first member of the pair is *x*, *y*, or *z* and the second member of the pair is 1 or 2. $A \times B = \{(x, 1), (x, 2), (y, 1), (y, 2), (z, 1), (z, 2)\}$. Note that the elements in the set which is the cartesian product are all ordered pairs and that there are six elements (pairs).

The cartesian product of a set *S* and itself ($S \times S$) contains all possible pairings of the elements in *S*. For example, if W is the set of whole numbers, W × W is the set

$$\{(0, 0), (0, 1), (0, 2) \ldots, (1, 0), (1, 1), (1, 2) \ldots, (2, 0), (2, 1), (2, 2) \ldots\}.$$

The cartesian product W × W is the infinite set of all ordered pairs of whole numbers.

For example, the cartesian product W × W (W is the set of whole numbers) is the set of all possible ordered pairs of whole numbers. To each of these pairs addition assigns exactly one sum and it is a whole number. Since addition is a function from W × W to W, we say that addition is a binary operation on the set of whole numbers.

Is addition also a binary operation on the integers? The question may be stated like this: If Z is the set of integers, is addition a function from Z × Z to Z? Is multiplication a binary operation on the whole numbers: i.e., is multiplication a function from W × W to W? Why?

Let us consider whether subtraction of whole numbers is a function. We can ask, "Is subtraction a binary operation on the set of whole numbers?" or "Is subtraction a function from W × W to W?" We must decide whether, for each ordered pair of whole numbers in W × W, subtraction assigns exactly one whole number. Examples of ordered pairs in W × W are

(6, 4),   (7, 1),   (3, 3),   (4, 6),   (7, 9),   (0, 5), ...

For each of the ordered pairs (6, 4), (7, 1), (3, 3), subtraction assigns exactly one whole number. But there is no whole number which subtraction assigns to the ordered pair (4, 6) because $4 - 6$ is not a whole number. Thus subtraction of whole numbers is not a function from W × W to W and subtraction is not an operation on the set of whole numbers.

Is there any set of numbers such that subtraction *is* an operation on that set? We agreed that $4 - 6$ does not name a whole number. But $4 - 6$ does name an integer. For each pair of integers, subtraction assigns exactly one integer; that is, subtraction is a function from Z × Z to Z. Subtraction is a binary operation on the set of integers.

$$(6, 4) \longrightarrow 2 \qquad (4, 6) \longrightarrow {}^-2 \qquad ({}^-1, {}^-1) \longrightarrow 0$$
$$(7, 7) \longrightarrow 0 \qquad ({}^-2, 6) \longrightarrow {}^-8 \qquad (5, 5) \longrightarrow 0$$
$$(7, 4) \longrightarrow 3 \qquad (4, 7) \longrightarrow {}^-3 \qquad ({}^-2, 0) \longrightarrow {}^-2$$

Thus binary operations on sets of numbers are examples of functions. To each element in a given set $S \times S$, a set of ordered pairs of numbers, an operation assigns exactly one element of the set S.

These binary operations are many-to-one functions because many different ordered pairs of numbers are assigned the same number. For example, multiplication assigns the same rational number to each of these ordered pairs: $(\frac{1}{2}, \frac{3}{2})$, $(2, \frac{3}{8})$, $(3, \frac{1}{4})$, $(\frac{1}{8}, 6)$. But to any ordered pair of rational numbers, exactly one rational number is assigned.

measurement Measurement establishes an assignment from certain sets of points (geometric figures such as line segments) to sets of numbers. For example, with respect to a given unit of measure, a positive number is assigned to each line segment. When line segments are measured, the language employed suggests that for the set A of all line segments and the set B of all positive real numbers:

Measurement (with respect to a given unit) assigns to *each* element of set A (each segment) *exactly one* element of B (a positive real number).

A similar assignment is established when we measure angles. In this case the unit of measure selected is usually a degree. To each angle, measurement assigns exactly one positive real number between 0 and 180. When we are measuring regions, such as a rectangular region, measurement assigns to the region exactly one real number, with respect to a square unit. It is the number of square units.

Thus measurement provides other examples of functions. To each element in a given set of geometric figures (segments, angles, rectangular regions, etc.) measurement assigns exactly one element of a set of numbers, always with respect to a given unit of measure.

A measurement function is many-to-one because many different figures will have assigned to them the same number, with respect to a given unit of measure. In the case of line segments this means that all congruent line segments have the same length. But to each line segment, exactly one length is assigned.

algebraic formulas Some open mathematical sentences suggest functions. For example, what is the *truth set* of the sentence

$$y = 2x + 4, \qquad x \text{ and } y \text{ whole numbers?}$$

The truth set is the set of ordered pairs that make the sentence true.† For this sentence:

If x is the whole number 0, then y is 4 because $4 = 2 \cdot 0 + 4$.

If x is 0, could we select a number for y other than 4 to make the sentence true?

If x is 1, then y is 6, because $6 = 2 \cdot 1 + 4$. No other number makes the sentence true where $x = 1$.

If x is 2, then y must be 8.

If x is 4, then y must be 12.

We see that given any whole number x, there is exactly one whole number y which makes the sentence true. Thus the algebraic formula $y = 2x + 4$ defines a function.

Does the open sentence $x = y^2$, x a whole number, define a function? Suppose that x is 4. Then y can be either 2 or $^-2$ because $4 = 2^2$ and $4 = (^-2)^2$ are true sentences. To the number x the sentence assigns more than one number y, and thus the sentence does not define a function from the set of numbers for x to the set of numbers for y.

4–3
definition of function

The common characteristic in all the examples given above is that there is assigned to each element in a set A exactly one element in a set B. Thus we shall define a function as follows:

> Given two nonempty sets A and B (not necessarily distinct), a function from A to B assigns to each element of set A exactly one element of set B.

A function is often denoted by a lower-case letter, such as f. If a is an element of set A and f assigns to a the element b in set B, this fact is variously denoted by:

$$f(a) = b \quad \text{or} \quad f: a \rightarrow b \quad \text{or} \quad a \xrightarrow{f} b.$$

---

† The truth set may also be called the *solution set*. It is the set of solutions of the sentence.

The notation is read "*f* of *a* is *b*" or "*f* at *a* is *b*" or "*f*-image of *a* is *b*." The element $f(a)$ in set B is called the *f*-image of element *a* in set A. In the example, *b* is the *f*-image of *a*.

Set A is called the *domain* of the function f. The set of all *f*-images of the elements of A is the *range* of the function f. Thus the range might consist of some but not all the elements of set B, or it might consist of all the elements of B. In either case, we say that the function is from A to B. But when a function has the property that the range consists of all elements in set B, then we can also say that the function is from A onto B. Such functions are called *onto* functions, since the set of images fills up or is onto the set B. For an onto function, no element of B is left unassigned to an element of A.

A function *f* from *A* to *B* is *onto* if for every element *b* of *B* there is an element *a* of *A* such that $f(a) = b$.

When a function has the property that each element of its range is the image of exactly one element of A, the function is called one-to-one (1–1). For 1–1 functions, no element of B is assigned to more than one element of A.

A function *f* from *A* to *B* is *one-to-one* (1–1) if for any elements *c, d* in *A*,

$$f(c) = f(d) \text{ implies that } c = d.$$

If there exists a function from A to B that is both 1–1 and onto, the two sets A and B are called equivalent. In such a case each element of B is the image of just one element of A.

A function from set *A* to set *B* that is 1–1 and onto is called a *one-to-one correspondence* between *A* and *B*.

Examples of four functions exhibiting various combinations of the properties of being 1–1 and/or onto are as follows:

Given: $A = \{a, b, c, d\}$;
$B = \{1, 2, 3, 4\}$;
$C = \{1, 2, 3\}$;
$D = \{1, 2, 3, 4, 5\}$

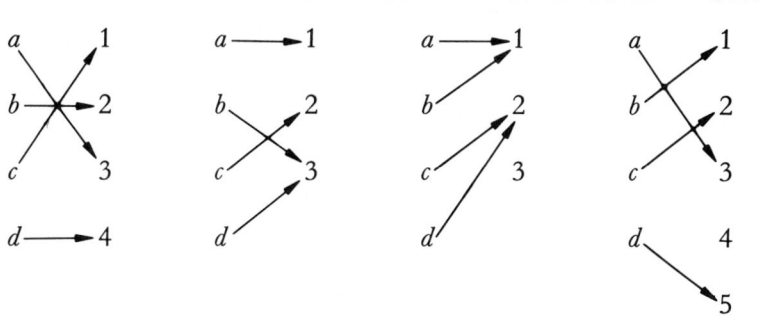

Set A $\xrightarrow{f}$ Set B   Set A $\xrightarrow{g}$ Set C   Set A $\xrightarrow{h}$ Set C   Set A $\xrightarrow{k}$ Set D

f is 1–1, and is onto. g is not 1–1, but is onto. h is not 1–1, and is not onto. k is 1–1, but is not onto.

The operation of addition of whole numbers is a function from the set of ordered pairs of whole numbers to the set of whole numbers. The addition function is onto, but not 1–1. Why? Subtraction is an operation on the set of integers. Is it onto?

The definition of a function and the examples given of functions make it clear that functions can be described as sets of ordered pairs. Consider two sets, the first five letters of the alphabet and the first five counting numbers, and a function f which assigns elements of the second set to elements of the first set in a way suggested by this set of ordered pairs:

$$f = \{(a, 1), (b, 2), (c, 3), (d, 4), (e, 5)\}.$$

The domain consists of the first members of the ordered pairs; the range consists of the second members of the ordered pairs. Thus, since our definition requires that a function assign *exactly one* element of the second set to *each* element of the first set, we can consider a function f as a set of ordered pairs in which no two distinct pairs have the same first member and in which every element of the domain of f occurs as a first member of a pair. The second member of the pair is the f-image of the first member.

exercise set 4–1

1. In each of the following, a set A and a set B are given, and an assignment is described. Decide whether the assignment is a function from A to B. If so, choose a few typical elements in A and determine the corresponding elements assigned in B. If a set is not a function, explain why not.

116 functions in geometry: motions in the plane

Set A	Set B	Name and description of assignment	Example
a) Integers	Integers	s: To each element of A assign its square	$s(3) = 9$ or $3 \xrightarrow{s} 9$
b) Whole numbers	Whole numbers	r: To each element of A assign its square root	$r(9) = 3$ or $9 \xrightarrow{r} 3$
c) Rectangles	Real numbers	a: To each element of A assign its area in square inches	$\begin{array}{c} 5 \\ 2\,\square \end{array} \xrightarrow{a} 10$
d) Events in the experiment of tossing 3 true coins	Rational numbers	p: To each element of A assign its probability	$p(3 \text{ heads}) = \frac{1}{8}$ or $3 \text{ heads} \xrightarrow{p} \frac{1}{8}$
e) Integers	Integers	f: To each x in A assign y in B such that $x + y = 3$ is true	$f(^{-}1) = 4$
f) Integers	Integers	g: To each x in A assign y in B such that $x + 2y = 3$ is true	$g(^{-}1) = 2$
g) Rational numbers	Rational numbers	q: To each x in A assign y in B such that $2x + y = 3$ is true	$q(^{-}\frac{1}{2}) = 4$

Set A	Set B	Name and description of assignment	Example
h) Ordered pairs of integers	Integers	m: To each element of A assign the product of the pair	$m(3, {}^-2) = {}^-6$ or $(3, {}^-2) \xrightarrow{m} {}^-6$
i) Ordered pairs of integers	Integers	d: To each element of A assign the quotient of the first by the second of the pair	$d(6, 2) = 3$
j) Ordered pairs of whole numbers	Whole numbers	v: To each element of A assign the average of the pair	$v(5, 7) = 6$
k) First-class letters	Positive integers	c: To each element of A assign the cost (in cents) of first-class postage	$3\frac{1}{2}$ oz. letter $\xrightarrow{c}$ 24
l) Students in a given class	Positive integers	h: To each element of A assign its height (in inches)	(John) $\xrightarrow{h}$ 38
m) Real numbers	Real numbers	w: To each x in A assign y in B such that $x = y^2$ is true	$w(4) = {}^-2$

2. a) For each of the assignments in Exercise 1 that is a function, determine its range. (For example, the range of function c in Exercise 1(k) is the set of all multiples of 6.)
   b) Which of these functions are *onto* set B, that is, have their ranges the same as their set B? (For example, function c is not onto the set of positive integers, since the range of c does not contain every positive integer.)

3. Which of the functions in Exercise 1 are 1–1? Explain why. (For example, function c in Exercise 1(k) is not 1–1 because two letters of different weights can have the same image under c:

$$(3\tfrac{1}{4} \text{ oz. letter}) \xrightarrow{c} 24, \qquad (3\tfrac{3}{4} \text{ oz. letter}) \xrightarrow{c} 24.)$$

4. a) Make a table of all the functions in Exercise 1, showing which are 1–1 and which are onto.
   b) Use the table to decide which functions are 1–1 correspondences.

5. How could Set B in Exercise 1(f) be changed so that the assignment g would then be a function from A to B? [*Hint:* Try B as the set of rational numbers.]

*6. Consider the functions s and f in Exercises 1(a) and (e). Let $f \circ s$ be the relation which assigns to each integer x an integer as follows:

$$f \circ s(x) = f[s(x)].$$

For example, $s(^-2) = 4$ and then $f(4) = ^-1$, so that $f \circ s(^-2) = f[s(^-2)] = f(4) = ^-1$.
   a) Determine $f \circ s(3), f \circ s(^-1), f \circ s(0)$.
   b) Describe the assignment $s \circ f$. Is it different from $f \circ s$? What is $s \circ f(3)$?

teaching questions and projects 4–1

1. Discuss the development of the idea of a function from examples to the generalized concept. At what point or level in a school program should the term and the general definition be introduced?

2. How could addition and multiplication be taught so as best to emphasize that they are functions?

3. Relate a child's discovery that any two numbers have just one sum to the definition of a function.

4-4
functions in geometry

It has been traditional to introduce functions explicitly for the first time in an algebra course. Unfortunately, the impression was often gained that a function is an algebraic formula and can have only sets of numbers for its domain and range. This is a very limited view of functions, for the power of the concept lies in its generality. It is applicable in all branches of mathematics, and this is why it is referred to as a unifying thread.

The sets on which we define a function may have numbers as elements, but may also have geometric points, or any other mathematical entities, as elements. The concept of a function is especially effective in understanding the theory of motions and congruences of geometric figures.

rigid motions In Chapters 1 and 3, you were asked to imagine that geometric figures move from one position to another by movements of the plane. But a geometric figure is a set of points, and we cannot really move a set of points. We can, however, make a copy of the figure, move the copy from place to place, and discuss the "apparent" motion of the figure.

Consider any line segment $\overline{XY}$ in the plane. Imagine that we move it to some other position in the plane, say to line segment $\overline{X'Y'}$. (It is as if we trace a drawing of $\overline{XY}$ and move the tracing to find that it now "fits" $\overline{X'Y'}$ exactly.) If $\overline{X'Y'}$ is congruent to $\overline{XY}$, and if the same movement takes every line segment of the plane to a congruent line segment, we call this a *rigid motion*. That is, rigid movements do not change lengths of line segments and, as a result, do not change sizes and shapes of figures as they are moved.

We can describe a rigid motion by a function. The two sets, A and B, are each the set of all points of the plane and the rigid motion is a one-to-one, onto function from A to B. It assigns to each point X a point X' (the image of X) and to each point Y a point Y' (the image of Y) such that

$$\overline{XY} \text{ is congruent to } \overline{X'Y'}$$

(in the sense of Postulate 2, Section 3–3). In terms of functional notation, this can be said as follows (see Fig. 4–5):

> **f is a *rigid motion* of the plane if and only if f is a 1–1, onto function such that for any two points X and Y,**
>
> $\overline{XY}$ **is congruent to** $\overline{f(X)f(Y)}$.

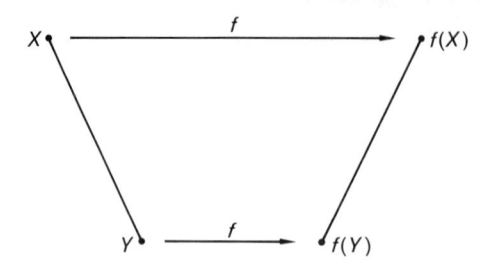

Figure 4–5

We sometimes refer to a function as a *mapping*. We say that a rigid motion *maps* every point onto its image. If the image of X is X', that is, $f(X) = X'$, then we say that X is mapped onto X'. If the image of $\overline{XY}$ is $\overline{X'Y'}$, then each point of $\overline{XY}$ is mapped onto exactly one point of $\overline{X'Y'}$.

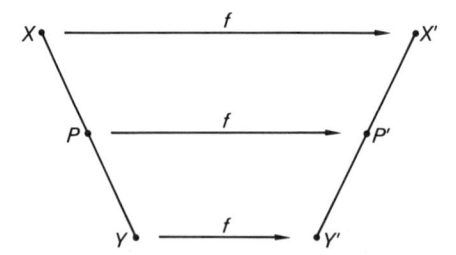

Figure 4–6

Since every function which is a rigid motion has as its domain all points of the plane, and since each point of its domain must have an image, the "apparent" motion of points under a rigid motion is not just the mapping of points in a particular geometric figure but a movement or mapping of *all* points of the plane. Every point of the plane has exactly one image, which is some point of the plane. Furthermore, a rigid motion maps line segments onto line segments. To understand this, let f be a rigid motion of the plane and P any point of $\overline{XY}$, and suppose that f

maps P onto P' (see Fig. 4–6). Then, since f is a rigid motion, $XP = X'P'$, $PY = P'Y'$, and $XY = X'Y'$. Also, since P is on $\overline{XY}$, we know that $XP + PY = XY$. This implies that $X'P' + P'Y' = X'Y'$, which tells us that P' is a point of $\overline{X'Y'}$. (Why?)

translations We have previously considered three rigid motions of the plane: parallel movements or translations, turning movements or rotations, and folding movements or reflections. A characteristic common to all these rigid motions, by definition, is that they are 1–1, onto mappings of the plane that preserve lengths of line segments. Each has other and distinctive properties.

Consider line segment $\overline{XY}$. We can imagine moving it to another position called $\overline{X'Y'}$ by a parallel movement, a translation, in which all points of the plane move the same distance and all along parallel paths. (See Fig. 4–7.)

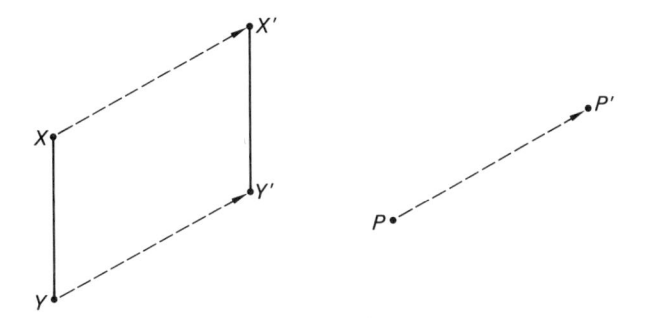

Figure 4–7

Let $\mathbb{P}$ be the set of points of the plane. Then a *translation* is a 1–1, onto mapping t from $\mathbb{P}$ to $\mathbb{P}$ such that for all points X, Y in $\mathbb{P}$, if

$$t(X) = X' \quad \text{and} \quad t(Y) = Y',$$

then

$$XX' = YY' \quad \text{and} \quad \overline{XX'} \parallel \overline{YY'}$$

and

$$\overline{XY} \parallel \overline{X'Y'};$$

consequently,

$$XY = X'Y'. \quad \text{(Why?)}$$

As with all rigid motions, the length of a segment is invariant under translations; that is, translations preserve length. The distinguishing characteristics of the rigid motion we call a "translation" are:

1) Every segment is parallel to its image; the distance from a point to its image is constant for all points; and paths from points to their images are parallel.
2) Except for the *identity* translation (in which every point is its own image), no point in the plane is a fixed point. (This means that no point is mapped onto itself or that no point is its own image.)

A translation is defined by a distance and a direction. (All points "move" the same distance and along parallel paths in the same direction.) We usually represent a translation with an arrow drawn from any point to its image (Fig. 4–8).

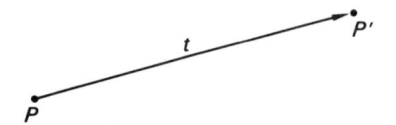

Figure 4–8

Let us consider a translation t that maps a geometric figure, say triangle ABC, onto its image, say triangle DEF. We associate with each point in $\triangle ABC$ exactly one point in $\triangle DEF$ (see Fig. 4–9). For example, $t(A) = D$, $t(B) = E$, and $t(C) = F$. Not only are the vertices of $\triangle ABC$ mapped onto the vertices of $\triangle DEF$, but each point of the triangle is associated with a corresponding point of its image. In fact, each point of the plane is mapped onto exactly one point of the plane, but we need only

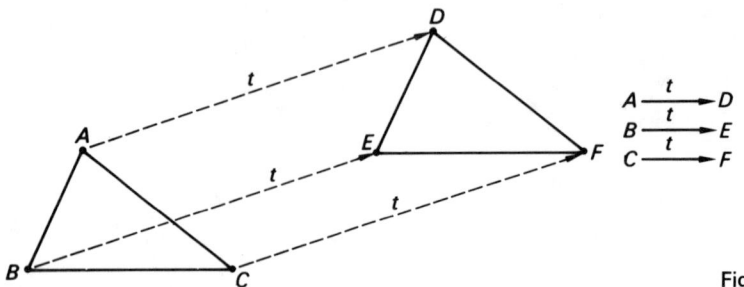

Figure 4–9

know the positions of one point and its image (for example, $A \xrightarrow{t} D$) to determine the "motion" of all other points of the plane under the translation.

Sliding a tracing from $\triangle ABC$ to $\triangle DEF$ illustrates that the triangles are the same size and shape, every segment is the same length as its image, and the measures of angles are preserved. We note also that $\overline{AB} \parallel \overline{DE}$, $\overline{BC} \parallel \overline{EF}$, and $\overline{AC} \parallel \overline{DF}$.

rotations Turning movements or rotations about a point are also mappings from the plane to the plane that preserve length. Imagine putting a pin through a piece of paper and turning the paper. One point remains fixed, and the others rotate along the arcs of circles. One point, X, is mapped onto itself; all other points of the plane are mapped onto other points except when the mapping is the identity (see Fig. 4–10).

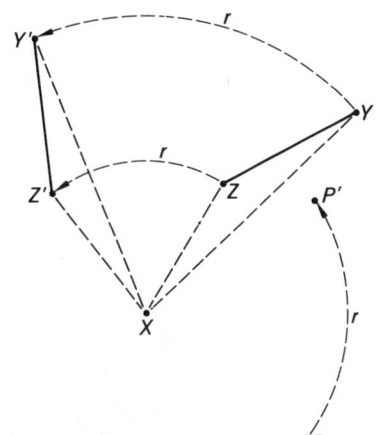

Figure 4–10

A *rotation* r (see Fig. 4–10) is a 1–1, onto mapping from $\mathbb{P}$ to $\mathbb{P}$ such that for any points X, Y, Z in $\mathbb{P}$, if

$$r(X) = X, \quad r(Y) = Y', \quad r(Z) = Z',$$

then

$$\text{angle } YXY' \equiv \text{angle } ZXZ', \quad XY = XY', \quad XZ = XZ',$$

and consequently,

$$YZ = Y'Z'. \quad \text{(Why?)}$$

A rotation r has the property of mapping each point P (other than X) of the plane $\mathbb{P}$ onto a point P' such that $XP = XP'$ and $\angle PXP' \equiv \angle YXY'$. $\angle YXY'$ is called the *directed angle of rotation*

124 functions in geometry: motions in the plane

from Y to its image Y', and point X is called the *center* or *fixed point* of the rotation. There are two possible directions for an angle of rotation: clockwise or counterclockwise. In Fig. 4–10 the directed angle of r is counterclockwise.

A rotation is determined by a fixed point and a directed angle of rotation. The directed angle of rotation from points to their images is constant for all points.

Suppose that the directed angle of rotation about point X is 360 degrees counterclockwise. Then each point is its own image because the "motion" is a full turn. A 360-degree rotation is the *identity* rotation.

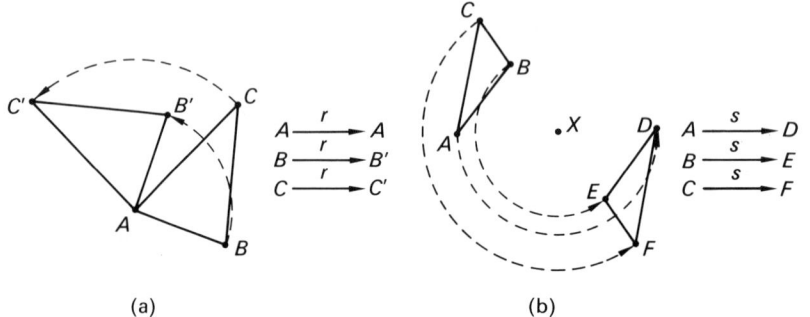

Figure 4–11

The fixed point, of course, may be any point of the plane. In Fig. 4–11(a) the fixed point is A, a vertex of $\triangle ABC$. The function r is a 90-degree rotation (a quarter turn) counterclockwise about point A mapping $\triangle ABC$ onto its image $\triangle AB'C'$. In Fig. 4–11(b) the fixed point is X and the function s is a 180-degree rotation (a half-turn) counterclockwise about point X mapping $\triangle ABC$ onto its image $\triangle DEF$.

In Fig. 4–11(a), $r(A) = A$ (A is mapped onto itself), $r(B) = B'$ (B is mapped onto B'), and $r(C) = C'$ (C is mapped onto C'). In Fig. 4–11(b), $s(A) = D$, $s(B) = E$, and $s(C) = F$.

For both rotations r and s illustrated, each segment has the same length as its image. The images of $\triangle ABC$ under each rotation are congruent to $\triangle ABC$. In Fig. 4–11(a), $\angle CAC' \equiv \angle BAB'$, $AC = AC'$, $AB = AB'$. Can you prove that $\triangle ABC \equiv \triangle AB'C'$? In Fig. 4–11(b), $\angle AXD \equiv \angle BXE \equiv \angle CXF$ and $AX = DX$, $BX = EX$ and $CX = FX$. Can you prove that $\triangle ABC \equiv \triangle DEF$?

reflections A third rigid motion of the plane, a folding movement or a reflection in a line, also preserves lengths of line segments. Any line in the plane determines a reflection. There are several ways to represent the idea intuitively (see Fig. 4–12). We can consider the line of reflection as the edge of a mirror so that the corresponding figure is the mirror image of the figure we have drawn. Try this, and draw the mirror image you see on the other side of the reflection line. Compare the two figures. We note, of course, that the figures are reversed. This is a familiar phenomenon. Letters of the alphabet are reversed in a mirror image. Cars driving on the right appear in a mirror to be on the left, and so on. But apart from this reversal, mirror images are not otherwise changed in size or shape.

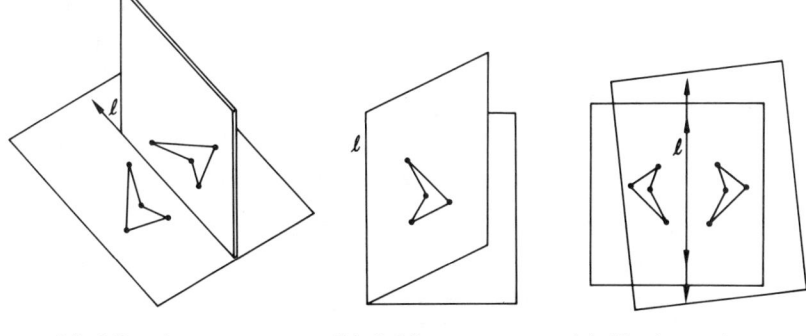

Figure 4–12   (a) Mirror image   (b) Folding paper   (c) Flip the tracing over

Another way to illustrate a reflection is to draw a figure on a piece of paper, fold along the line of reflection and trace the image on the other side of the fold. Or we can trace the figure and the line of reflection, then turn the tracing paper over so that the traced line again falls on the line of reflection, and then draw the image.

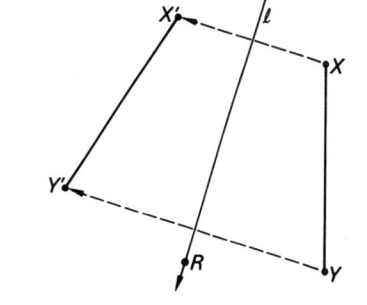

Figure 4–13

> A *reflection* f in a line l is a 1–1, onto mapping from $\mathbb{P}$ to $\mathbb{P}$ such that each point of l is left fixed by the mapping and if X does not belong to l and $f(X) = X'$, then l is the perpendicular bisector of $\overline{XX'}$ (Fig. 4–13). As a consequence, for any X, Y in $\mathbb{P}$, if $f(X) = X'$ and $f(Y) = Y'$, then $XY = X'Y'$ (why?).

Reflections preserve lengths of line segments. Thus they are rigid motions. The distinguishing properties of reflections are:

1) The reflection line is the perpendicular bisector of each segment from a point to its image.
2) Each point on the line of reflection is a fixed point of the reflection.

A reflection is completely determined by its line of reflection, and the points of the line are the only fixed points of the reflection.

The image of any figure under a reflection about a line is the same "size and shape" as the figure. This is an intuitive way of comparing two figures: do they fit each other exactly? In fact, as with all rigid motions, a figure and its image under a reflection are congruent.

congruence and rigid motions The concept of function enables us to describe mathematically the idea that is suggested to our intuition by "motions" of figures. A function describes precisely and explicitly the apparent movement of each point of the plane to some other point by relating these points in pairs, such that to each point of the plane exactly one point is assigned by the "motion."

When we trace figures and when we slide, turn, or flip the tracing over to find an image of the figure, it is intuitively obvious that the figure and its image are the same size and shape. Roughly, our initial ideas of congruent figures meant figures "of the same size and shape" or figures "that fit each other exactly."

Thus the function concept, when used to describe rigid motions, gives an insight into the idea of congruence. By definition, a rigid motion is a motion of the plane that preserves lengths of segments. When we understand the function concept, we understand that when we say that two figures are "the same size and shape," we mean that there is a rigid motion of the plane, a 1–1, onto, distance-preserving function from $\mathbb{P}$ to $\mathbb{P}$, that maps one figure onto the other. Thus congruence is a one-to-one correspondence of points of the plane that preserves distance.

Translations, rotations, and reflections are all cases of rigid motions of the plane. We shall agree that

Two figures in a plane are *congruent* if and only if there is a rigid motion that maps one of the figures onto the other.

exercise set 4–2

1. Use the properties of a rigid motion to construct (if possible) a square with a vertex on each of the circles and two vertices on line k. [*Hint:* Consider the intersection of one circle with the reflection of the other in k.]

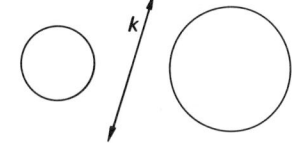

2. Use the properties of a rigid motion to help you find the shortest path from point A to a point on line m to point B. [*Hint:* Use the reflection of B in line m.]

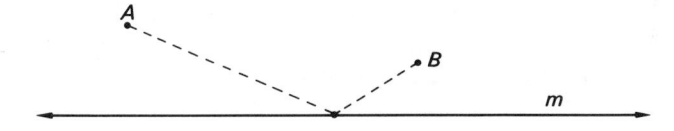

3. Given a triangle ABC with $\overline{AB} \equiv \overline{CB}$. Use the properties of rigid motions to show that $\angle BAC \equiv \angle BCA$.

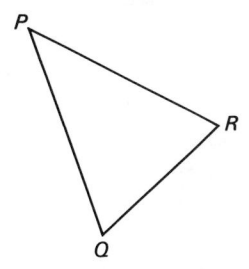

4. Show by using a rigid motion that the sum of the measures of the angles of a triangle PQR is equal to the measure of a straight angle.

5. *ABCD* is a square.
   a) Describe the rigid motion that maps $\triangle ABC$ onto $\triangle CDA$ such that: $A \rightarrow C, B \rightarrow D, C \rightarrow A$.
   b) Describe the rigid motion that maps $\triangle ABC$ onto $\triangle ADC$ such that: $A \rightarrow A, C \rightarrow C, B \rightarrow D$.

6. Describe the fixed points under the following rigid motions: (a) a translation, (b) a rotation, (c) a reflection, (d) the identity function.

7. a) What type of motions will show the congruence of two right-hand glove patterns of the same size?
   b) What types of motions will show the congruence of a right- and a left-hand glove pattern of the same size?

teaching questions and projects 4-2

1. Plan an activity in which children describe rigid motions that transform letters or numerals or pictures into other symbols or pictures. For example:
   a) N Z
   d) d p
   g)

   b) M W
   e) d b

   c) b p
   f) 1889 6881

2. Discuss the development from activities using tracing paper and/or folding paper to the notion of rigid motions as one-to-one functions. How does the intuitive idea of congruence as "fitting exactly" lead to a more precise idea of congruence?

4-5
composition of functions

When we move a copy of a figure to fit another figure we usually carry out several movements. For example, we may slide, then turn a figure until it fits, or slide, turn, and then flip the copy over. Such activities may be described as a series of rigid motions.

Imagine that we trace a copy of a geometric figure, then slide it so that its points follow parallel paths to a new position, then slide it along parallel paths again to a third position. Is there a motion that would take the figure from its original position directly to the third position? For example, let t_1 and t_2 be two translations of the plane. If t_1 is a function that maps point P onto Q and if t_2 is a function that maps point Q onto R, then the translation t_1 followed by the translation t_2 has the resulting effect of mapping point P onto R. We use the notation $t_2 \circ t_1$† to denote the motion of the plane that maps point P *directly* onto R. (See Fig. 4–14.)

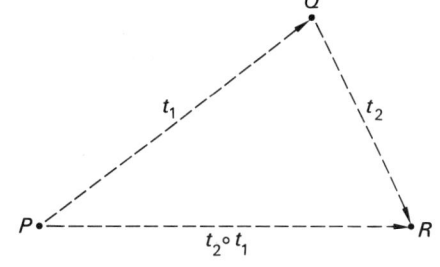

Figure 4–14

The translations t_1 and t_2 are functions of the plane; hence the motion $t_2 \circ t_1$ is also a function of the plane, since it assigns to each point of the plane exactly one point of the plane. In functional notation:

If $t_1(P) = Q$ and $t_2(Q) = R$,

then

$$t_2 \circ t_1(P) = t_2[t_1(P)] = t_2(Q) = R.$$

Note that we write the symbol for the composite function from right to left to show that it is a *function of a function*. We first find the image of P under t_1, which is Q. Then we find the image of Q under t_2. It is R. The composite function $t_2 \circ t_1$ represents t_1 followed by t_2. So $t_2 \circ t_1(P) = R$.

This example leads us to a general definition of the composite of two functions.

> If f is a function from A to B and g is a function from the range of f (a subset of B) to C, then $g \circ f$ is a function from A to C defined for all a in A by $g \circ f(a) = g(f(a))$. The function $g \circ f$ is called the *composite* of f and g.

---

† It is sometimes instructive to denote the composite of two functions by placing a small circle between them, to remind us that $z \circ m$ is not a product. We shall use the notation zm when it is clear that we are referring to a composite.

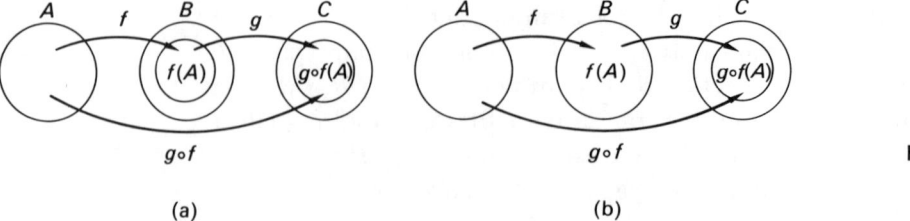

Figure 4–15

(a)

(b)

The composite function $g \circ f$, where the range of f is a proper subset of set B, is illustrated in Fig. 4–15(a). The function f is not an onto function in this case, and the domain of function g does not contain all elements of B, but only that proper subset of B which is the range of f. Thus g is a function from a proper subset of B to C (not necessarily from B to C).

Figure 4–15(b) illustrates the case of composite function $g \circ f$ where f is an onto function. In this case the range of f is also a subset of B, namely the set B itself. We say in this case that g is a function from B to C.

Composite functions are not confined to geometry. The idea can also be illustrated using algebraic functions. For example, consider two functions f and g from the integers, defined by the equations

$$f(x) = x + 1, \qquad g(x) = 2x.$$

Thus function f has the effect of adding 1 to each integer; function g has the effect of doubling each integer. The composite function $g \circ f$ is defined by

$$g \circ f(x) = g(f(x)) = g(x + 1) = 2(x + 1).$$

Suppose that x is 5. Then $g \circ f(5) = g(f(5)) = g(5 + 1) = 2(5 + 1)$. So $g \circ f(5) = 12$. The image of 5 under f is $5 + 1$ or 6. Then the image of 6 under g is $2(6) = 12$. If x is $^-3$, what is $g \circ f(^-3)$?

Does the order of the functions in the composite affect the result? For example, is $g \circ f(x) = f \circ g(x)$ for all x? We found that $g \circ f(x) = 2(x + 1)$. But

$$f \circ g(x) = f(g(x)) = f(2x) = 2x + 1.$$

Since, for some x,

$$2(x + 1) \neq (2x) + 1,$$

we have shown that $g \circ f$ and $f \circ g$ are different functions. Thus we have demonstrated that *composition of functions is not commutative.*

Example: Given a function f_1 such that $f_1(x) = x + 3$, find a function f_2, if possible, such that

$$f_2 \circ f_1(x) = x.$$

In other words, find a function f_2 such that the composite $f_2 \circ f_1$ is the identity function (leaves all integers fixed).

Let $f_1(x) = y$. Then $y = x + 3$, so that $x = y - 3$. If f_2 exists such that $f_2(y) = x$, then $x = y - 3 = f_2(y)$. Thus f_2 assigns to each number y the number $y - 3$. As a check:

$$f_2 \circ f_1(x) = f_2(f_1(x)) = f_2(x + 3) = (x + 3) - 3 = x.$$

4–6
composition of rigid motions

The idea of composite functions raises some natural questions. Is the composite of two rigid motions also a rigid motion? Suppose that f and g are two rigid motions—that is, distance-preserving functions—of the plane. Since f is distance-preserving, it maps each segment $\overline{XY}$ of $\mathbb{P}$ onto a congruent segment $\overline{X'Y'}$. Also, since g is distance-preserving, it maps segment $\overline{X'Y'}$ onto a congruent segment $\overline{X''Y''}$. But then $\overline{X''Y''}$ must be congruent to $\overline{XY}$ (because they are both congruent to $\overline{X'Y'}$) and it follows that the composite function $g \circ f$ maps each segment $\overline{XY}$ onto a congruent segment $\overline{X''Y''}$. Therefore $g \circ f$ is a rigid motion whenever f and g are rigid motions.

The composite of any two rigid motions is a rigid motion.

This is a significant property of composition of rigid motions, for it implies that to each ordered pair of rigid motions composition assigns exactly one rigid motion. This means that:

Composition is an operation on the set of rigid motions.

By sliding copies of figures on a plane, we verify intuitively that the composite of two translations is a translation (see Fig. 4–14) and that composition is an operation on the set of translations. This will be proved in Chapter 5.

It is also intuitively clear that if r_1 and r_2 are rotations about the same center X, then the composite function $r_2 \circ r_1$ is also a rotation about X (see Fig. 4–16). The directed angle of the composite $r_2 \circ r_1$ has a measure which is the sum of the measures of the directed angles of rotation of r_1 and r_2. (We agree to measure counterclockwise directed angles with positive numbers and clockwise directed angles with negative numbers.)

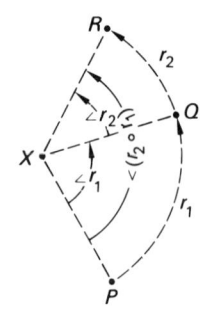

(a) $m(\angle r_2 \circ r_1) = m(\angle r_1) + m(\angle r_2)$
r_1 and r_2 both counterclockwise

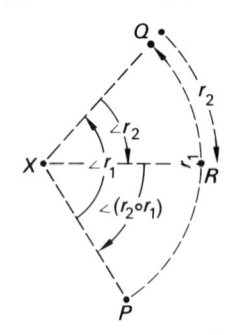

(b) $m(\angle r_2 \circ r_1) = m(\angle r_1) + m(\angle r_2)$
r_1 counterclockwise; r_2 clockwise;
$m(\angle r_1)$ is positive; $m(\angle r_2)$ is negative

Figure 4–16

properties of composition Let us examine the operation of composition of rigid motions to determine some of its properties. When we discussed an example of algebraic functions, we found that composition of functions is not commutative. In fact, composition of rigid motions is also *not commutative*, as we shall see later. But composition of rigid motions is *associative;* that is, for all rigid motions f_1, f_2, and f_3,

$$(f_3 \circ f_2) \circ f_1(P) = f_3 \circ (f_2 \circ f_1)(P).$$

It is meaningful to speak of two different translations or two different rotations about a point. A translation other than the identity will take any point to another point at the same distance and in the same direction from its initial point; there are infinitely many choices of distance and direction. A rotation other than the identity about a fixed center X will take any point (other than X) to another point depending on the directed angle of the rotation; there are infinitely many choices of angle. But we discover that a reflection about a line l is uniquely determined by the line l, so that there is only one possible reflection about l. The question, "What is the composite of two reflections in l?" must be rephrased, "What is the composite of the reflection in l with itself?" The composite is the function of the plane that takes each point of the plane onto itself. We have called this the *identity function*, the function that leaves each point of the plane fixed. Let f be the reflection in l. Then if $f(P) = Q$, it must also be true that $f(Q) = P$, so that

$$f \circ f(P) = f(f(P)) = f(Q) = P.$$

Since, for any point P, $f \circ f(P) = P$, the function $f \circ f$ is the identity function (Fig. 4–17). And since $f \circ f$ is the identity function, we say that f is the *inverse* of f. Thus every reflection about a line is its own inverse.

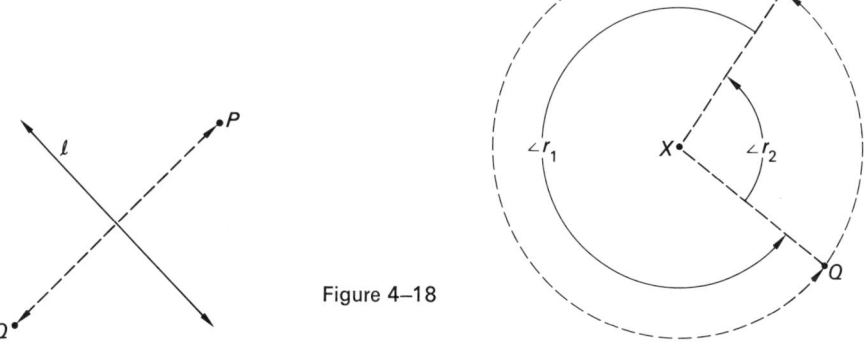

Figure 4–17

Figure 4–18

Is there a rotation which is the identity function? Is there a rotation that leaves every point of $\mathbb{P}$ fixed? A full turn about a given point (a rotation with an angle of 360 degrees) maps every point onto itself so that each point is its own image. Thus a full turn is the *identity* function. If r is a 360-degree rotation about a point, then for any point P, $r(P) = P$. Also a 0-degree rotation is the identity function. (Why?)

Given any rotation r_1 about a point X, is there another rotation r_2 such that $r_2 \circ r_1(P) = P$ and $r_1 \circ r_2(P) = P$? Intuitively, the question is: If we "turn" the plane through a certain directed angle, is there another "turn" we can make so that each point is taken back to itself? It should be clear that there is. For example, if r_1 is the rotation of 270 degrees counterclockwise about point X (see Fig. 4–18), then the rotation r_2 of 90 degrees counterclockwise

134 functions in geometry: motions in the plane

about point X is a rotation such that composite $r_2 \circ r_1$ takes every point P to itself. We say that r_2 is an *inverse* of r_1 because $r_2 \circ r_1$ is the identity. Thus every rotation has an inverse rotation.

other compositions There are many other possible compositions of rigid motions in which we consider composites of rotations about different centers, reflections about different lines, or combinations of these. For example, consider the reflection f_1 in line l_1, and the reflection f_2 in line l_2. Is $f_2 \circ f_1$ a reflection, and if so, in what line? There are two cases to be considered: either $l_1 \parallel l_2$ or l_1 intersects l_2 in a point Z.

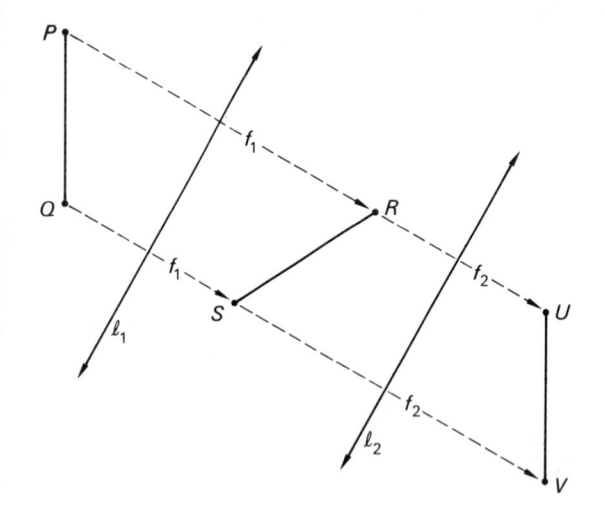

Figure 4–19

If $l_1 \parallel l_2$, we can fold tracing paper at l_1, and then at l_2, to see what happens to a line segment $\overline{PQ}$ under the composite function $f_2 \circ f_1$ (see Fig. 4–19). We find that $f_1(\overline{PQ}) = \overline{RS}$ and $f_2(\overline{RS}) = \overline{UV}$, so that $f_2 \circ f_1(\overline{PQ}) = \overline{UV}$. Paper folding suggests that $f_2 \circ f_1$ is not a reflection; instead, it is a translation, since $\overline{PQ} \parallel \overline{UV}$. The same problem can be solved by constructing the segment $\overline{UV}$ from the given segment $\overline{PQ}$ with ruler and compass, and then proving that $\overline{PQ} \parallel \overline{UV}$. [*Hint:* Prove that $\overline{PU} \parallel \overline{QV}$ and $PU = QV$.] It can also be proved that the resulting translation is in direction perpendicular to l_1, and a distance twice the distance between l_1 and l_2. Also we can show that each translation is the composite of two reflections in parallel lines. The proofs are left as exercises.

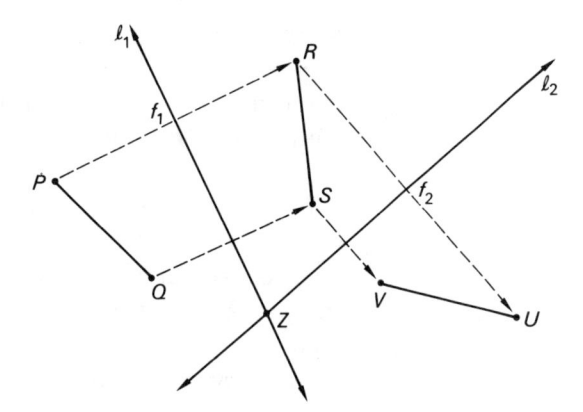

Figure 4–20

Now suppose that l_1 intersects l_2 at point Z. We can again fold tracing paper at l_1 and then at l_2 to see where line segment $\overline{PQ}$ is taken by the composite function $f_2 \circ f_1$ (see Fig. 4–20). Paper-folding suggests that $f_2 \circ f_1$ is actually a rotation about the point Z, the intersection of l_1 and l_2.

The construction of $\overline{UV}$ from $\overline{PQ}$ with ruler and compass in this case is an interesting problem, and the related proof that $\angle QZP \equiv \angle VZU$ is a good example of proofs using congruence. [*Hint:* Show that $\triangle QZP$ and $\triangle VZU$ are congruent to $\triangle SZR$.] We can also prove that the angle of rotation has twice the measure of the angle formed by l_1 and l_2. Reversing this construction, we see that each rotation is the composition of two reflections in lines intersecting in the center of rotation. The proofs are left as exercises.

It is clear that the composite of two translations is a translation. But some cases of composition of rigid motions are not as easy to recognize as others. Experiment with various composites to answer questions such as: Is the composite of two rotations with different centers a rotation? What is the composite of a rotation and a translation, a reflection and a rotation? You will find, for example, that the composite of two rotations with different centers is usually a rotation with a third center (it is an interesting problem to construct the center of the composite rotation); that the composite of a translation and a rotation is a rotation about a new center; and that any rotation is the composite of two reflections and that any translation is the composite of two reflections. Thus any geometric figure can be mapped onto any congruent figure by a series of reflections alone.

glide reflection The rigid motions can be characterized by their fixed points. For example, a rotation (other than the identity) has exactly one fixed point, and conversely, every rigid motion with exactly one fixed point is a rotation about the point. Also, a reflection has a fixed line (each point of which is a fixed point), and conversely, every rigid motion that leaves the points of a line fixed, and only these points, is a reflection in that line. The question remains: If a rigid motion has no fixed points, is it a translation? Not necessarily. Although a translation has no fixed points (unless it is the identity), there is another type of rigid motion, called a glide reflection, that also has no fixed points.

> A *glide reflection* is a composite of a reflection about a line l and a translation parallel to l. A function g is a glide reflection about l if for any point P in the plane, $g(P) = Q$, where Q is obtained by reflecting P onto P' about l and then translating P' onto Q by a translation parallel to l (see Fig. 4–21).

The same is obtained by a translation parallel to l and then a reflection in l.

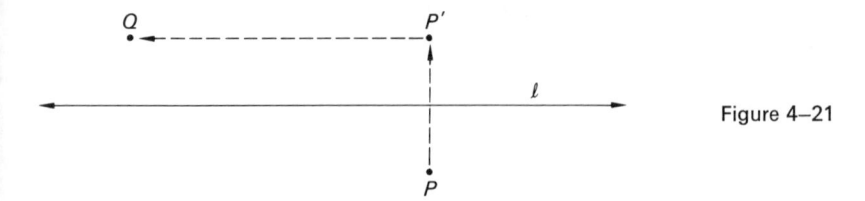

Figure 4–21

summary It can be shown that these four rigid motions—translations, rotations, reflections, and glide reflections—are the *only* rigid motions of the plane, that is, the only functions that preserve lengths of line segments and therefore congruence of figures. Since each translation is the composite of two reflections, each rotation is the composite of two reflections, and each glide reflection is the composite of three reflections (why?), we conclude that every rigid motion of the plane is the composite of at most three reflections. We saw that composition is an operation on the set of rigid motions. Another way of saying this is:

The composite of any two rigid motions is again a rigid motion.

If two plane figures are congruent, then one figure can be moved to the other by a rigid motion. In other words, if one figure is congruent to another, there is a function which is a rigid motion of the plane that maps the points of one figure onto the points of the other.

exercise set 4–3

1. a) Draw a line $\overleftrightarrow{EF}$. Make $\overline{EF}$ about 6 cm long. Draw any $\triangle RST$. By drawing an arrow, show a translation t of the plane such that $t(E) = F$.
   b) On the figure find $t(R)$, $t(S)$, $t(T)$, and draw the t-image of $\triangle RST$.
   c) Verify that $\overline{RS} \equiv \overline{R'S'}$ and $\overline{RS} \parallel \overline{R'S'}$, $\overline{ST} \equiv \overline{S'T'}$ and $\overline{ST} \parallel \overline{S'T'}$, $\overline{RT} \equiv \overline{R'T'}$ and $\overline{RT} \parallel \overline{R'T'}$, where $t(R) = R'$, $t(S) = S'$, $t(T) = T'$.

2. a) Draw a point X, a point A, and any $\triangle MTG$. Function r is a rotation of 60° clockwise about X. By drawing a directed arc of a circle, find the point A' so that $r(A) = A'$.
   b) Find the r-image of M, of T, of G, of $\triangle MTG$.
   c) Verify that $XM = XM'$, $XT = XT'$, $XG = XG'$; that $MT = M'T'$, $TG = T'G'$, $MG = M'G'$; that $\triangle MTG \equiv \triangle M'T'G'$, where $r(M) = M'$, $r(T) = T'$, $r(G) = G'$.

3. a) Draw a line l, a point Q, and a $\triangle DEF$. Let f be the reflection of $\mathbb{P}$ in l and find Q' so that $f(Q) = Q'$.
   b) Show the f-image of D, of E, of F, of $\triangle DEF$.
   c) Verify that $\overline{QQ'} \parallel \overline{DD'} \parallel \overline{EE'} \parallel \overline{FF'}$; $DE = D'E'$, $EF = E'F'$, and $DF = D'F'$, where $f(D) = D'$, $f(E) = E'$, $f(F) = F'$.

4. Let f_1 and f_2 be reflections of the plane in lines l_1 and l_2, respectively.
   a) Draw two parallel lines l_1 and l_2 and a line segment $\overline{PQ}$. Construct $\overline{RS}$, where $f_1(P) = R$, $f_1(Q) = S$. Then construct $\overline{UV}$, where $f_2(R) = U$, $f_2(S) = V$. Use your drawing to suggest a proof that $f_2 \circ f_1$ is a translation if l_1 and l_2 are parallel.
   b) Draw two lines l_1 and l_2 intersecting in a point Z, and a line segment $\overline{PQ}$. Follow the same instructions as in part (a). Measure the drawing and verify that $ZQ = ZV$, $ZP = ZU$, $\angle QZV \equiv \angle PZU$, and the measure of $\angle QZV$ is twice that of the angle from l_1 to l_2. Use the drawing to suggest a proof that $f_2 \circ f_1$

is a rotation about Z if l_1 and l_2 intersect in Z. Also prove that the measure of the angle of rotation is twice the measure of the angle formed by l_1 to l_2.

5. a) Mark two points X_1 and X_2 as centers of rotations r_1 and r_2 of 50° and 70° clockwise, respectively. Draw a line segment $\overline{PQ}$ and construct $\overline{P'Q'}$, where $r_1(P) = P'$, $r_1(Q) = Q'$. Then construct $\overline{P''Q''}$, where $r_2(P') = P''$, $r_2(Q') = Q''$. Thus $r_2 \circ r_1$ takes $\overline{PQ}$ to $\overline{P''Q''}$.
    b) Show that $r_2 \circ r_1$ is a rotation, as follows: Draw $\overline{PP''}$ and $\overline{QQ''}$, find their midpoints, and construct perpendicular bisectors to $\overline{PP''}$ and $\overline{QQ''}$. Let O be the point of intersection of these perpendicular bisectors. Measure your drawing to verify that O is the center of rotation of 50° + 70° which takes $\overline{PQ}$ to $\overline{P''Q''}$.

6. a) Draw any segment $\overline{PQ}$ and another segment $\overline{P'Q'}$ so that $\overline{P'Q'}$ is the image of $\overline{PQ}$ under a translation t. Choose a point X as the center of a clockwise rotation r of 60°. Construct $\overline{P''Q''}$, where $r(P') = P''$, $r(Q') = Q''$.
    b) Show that $r \circ t$ is a rotation as follows: Draw $\overline{PP''}$ and $\overline{QQ''}$ and construct their perpendicular bisectors. By measurements verify that the point of intersection of the perpendicular bisectors is the center of a rotation that takes $\overline{PQ}$ to $\overline{P''Q''}$. What is the measure of the angle of this rotation?

7. a) Follow the instructions of Exercise 5 with r_1 and r_2 rotations of 70° clockwise and 70° counterclockwise, respectively.
    b) Verify that $r_2 \circ r_1$ is a translation.

8. a) Follow the instructions of Exercise 6 with r a rotation of 180°.
    b) Verify that $r \circ t$ is also a rotation of 180°, and find its center.

9. a) Draw a line l and a line segment $\overline{PQ}$. Reflect $\overline{PQ}$ in l to $\overline{P'Q'}$ and translate $\overline{P'Q'}$ parallel to l any distance to $\overline{P''Q''}$. We say that $\overline{PQ}$ is taken to $\overline{P''Q''}$ by a glide reflection about l. Let $\overline{P''Q''}$ be taken by another glide reflection about l to $\overline{RS}$.
    b) On your drawing verify that $\overline{PQ} \parallel \overline{RS}$ and $PQ = RS$, that is, that one glide reflection about l followed by another about l is a translation.
    c) Is the translation found in (b) parallel to l? If so, prove it.

10. a) Follow the instructions of Exercise 9; however, let the second glide reflection be about a different line from the first, but parallel to the first.
    b) Verify on your drawing that the composite is again a translation.

11. With a series of drawings, illustrate a technique for:
    a) Replacing a given translation by a composite of two reflections.
    b) Replacing a given rotation by a composite of two reflections.
    [*Hint:* Refer to Exercise 4.]
12. With a series of drawings, illustrate the fact that:
    a) The inverse of a translation is a translation.
    b) The inverse of a glide reflection is a glide reflection.
13. What is the composition of two glide reflections about intersecting lines l_3 and l_4?
14. Give an argument that every glide reflection about a line l is the composite of three reflections. Describe the three lines of reflection.
15. Given a glide reflection g_1 about line l, describe another glide reflection g_2 such that $g_2 \circ g_1(P) = P$ for all points P of $\mathbb{P}$.

4-7
the group of rigid motions

A group is a mathematical structure found in many branches of mathematics and applicable outside mathematics as well. A group consists of a set S and a binary operation or function from $S \times S$ to S,* such that:

1) The operation is associative.
2) There is an identity element in S for the operation.
3) Every element in S has an inverse in S for the operation.

A familiar arithmetic example of a group is the set of integers under addition. Addition in $\mathbb{Z}$ (the set of integers) is an operation† from $\mathbb{Z} \times \mathbb{Z}$ to $\mathbb{Z}$, such that for all a, b, c in $\mathbb{Z}$,

1) $(a + b) + c = a + (b + c)$.
2) $a + 0 = a$ and $0 + a = a$.
3) For each integer a, there exists an integer ^-a (the opposite of a) such that $a + {}^-a = 0$ and $^-a + a = 0$.

Since addition in $\mathbb{Z}$ is commutative, we further describe this as a *commutative group*.

---

* Since the operation is a function from $S \times S$ to S, this is sometimes described by saying that the set is *closed* under the operation. Closure is a necessary property for groups.

† The set of integers is closed under addition.

On the other hand, the set of integers with the binary operation of multiplication is *not* a group because some integers do not have an inverse under multiplication.

The same structure is found in geometry. Consider R, the set of all rigid motions of $\mathbb{P}$, and the operation* of composition of rigid motions. If f, g, k are any rigid motions, then:

1) $f \circ (g \circ k) = (f \circ g) \circ k$. (Composition is *associative* in the sense that the composite functions $f \circ (g \circ k)$ and $(f \circ g) \circ k$ have the same effect on all points of the plane.)

2) There is a rigid motion I (the *identity function* that leaves every point of the plane fixed) such that $f \circ I = f$ and $I \circ f = f$ for all rigid motions f.

3) For every rigid motion f there is a rigid motion f^{-1} (called the *inverse* of f) such that $f \circ f^{-1} = I$ and $f^{-1} \circ f = I$ (where I is the identity function).

We see then that:

R, the set of all rigid motions of $\mathbb{P}$, and the operation of composition of rigid motions form a group.

4-8
symmetries as functions

In Chapter 1 we thought of a *symmetry of a figure* as a rigid motion of the plane that takes the figure into itself. Using this idea, we classified triangles as isosceles, equilateral, or neither, depending on the number of folding or turning symmetries of the triangle.

the group of symmetries of an equilateral triangle The concept of a group is probably most clearly illustrated in geometry with sets of symmetries. In Section 4-7 we saw that the set of all rigid motions is a group under composition. If we restrict our attention to those rigid motions of $\mathbb{P}$ that are symmetries of a simple figure, such as an equilateral triangle, we have a finite subset of the rigid motions that we shall show is itself a group.

Consider the equilateral triangle ABC and determine all its symmetries. The most obvious symmetry is the identity function I,

---

*The set of rigid motions is closed under composition.

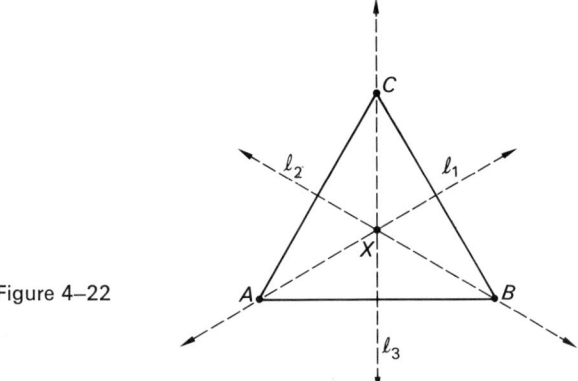

Figure 4–22

the full 360° rotation, which leaves all the points of the triangle unchanged. Each point is mapped onto itself by the identity. There are three folding symmetries (reflections) in the lines l_1, l_2, l_3; call them f_1, f_2, f_3, respectively (see Fig. 4–22). Thus

$$f_1(A) = A, \quad f_1(B) = C, \quad f_1(C) = B;$$
$$f_2(A) = C, \quad f_2(B) = B, \quad f_2(C) = A;$$
$$f_3(A) = B, \quad f_3(B) = A, \quad f_3(C) = C.$$

We have interpreted a rigid motion as a particular function whose domain is the set of all points of the plane. But for purposes of discussing the symmetries of a particular figure, we can simplify matters by restricting our domain to just the points included in the figure. In the case of the equilateral triangle, for example, it will be sufficient to limit the domain to just the vertices, A, B, and C, since the symmetries of the figure are completely determined by what the symmetries do to the vertices. Thus if the range of the function is the same set of points, A, B, and C, the function is a symmetry of the triangle. Each symmetry maps the figure onto itself. We can now indicate symmetries by the following notation:

$$f_1 = \begin{pmatrix} A & B & C \\ A & C & B \end{pmatrix}, \quad f_2 = \begin{pmatrix} A & B & C \\ C & B & A \end{pmatrix},$$

$$f_3 = \begin{pmatrix} A & B & C \\ B & A & C \end{pmatrix}, \quad I = \begin{pmatrix} A & B & C \\ A & B & C \end{pmatrix},$$

where the notation shows the image of a point directly below the point.

functions in geometry: motions in the plane

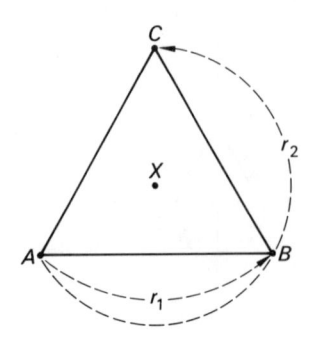

Figure 4–23

In addition to the identity function, which is a 360-degree symmetry of rotation, there are two other symmetries of rotation about the point X, where l_1, l_2, l_3 intersect. Let r_1 be a rotation of a one-third (120-degree) turn counterclockwise about X, and let r_2 be a rotation of a two-thirds (240-degree) turn counterclockwise about X (Fig. 4–23). Since $r_1(A) = B$, $r_1(B) = C$, $r_1(C) = A$, etc., we can write*

$$r_1 = \begin{pmatrix} A & B & C \\ B & C & A \end{pmatrix} \quad \text{and} \quad r_2 = \begin{pmatrix} A & B & C \\ C & A & B \end{pmatrix}.$$

Thus there are six symmetries of the equilateral triangle: three reflections and three rotations.

Let us first consider the set of three rotational symmetries of the equilateral triangle. These are the one-third-turn counterclockwise about X (r_1), the two-thirds-turn counterclockwise about X (r_2), and the full turn about X (the identity function I).

We shall first determine the composite function $r_2 \circ r_1$ by finding the image of each vertex under $r_2 \circ r_1$. For example,

$$r_1(A) = B \quad \text{and} \quad r_2(B) = A.$$

So

$$r_2 \circ r_1(A) = r_2[r_1(A)] = r_2(B) = A.$$

Also

$$r_2 \circ r_1(B) = r_2[r_1(B)] = r_2(C) = B$$

and

$$r_2 \circ r_1(C) = r_2[r_1(C)] = r_2(A) = C.$$

---

* Although we can say that the figure is an invariant set under a symmetry because it is its own image, it is not necessarily true that each point is fixed. For example, under r_1 no point of the triangle is fixed, although the triangle itself is an invariant set. In fact, only under I is *every* point fixed.

We represent the composite with the notation

$$r_2 \circ r_1 = \begin{pmatrix} A & B & C \\ A & B & C \end{pmatrix}$$

and this is the symmetry I; $r_2 \circ r_1 = I$.

Similarly we find that

$$r_1 \circ r_1 = \begin{pmatrix} A & B & C \\ C & A & B \end{pmatrix},$$

which is the symmetry r_2; $r_1 \circ r_1 = r_2$; etc.

We can list the composites in a chart. The composite of a symmetry shown in the left column followed by a symmetry shown in the top row is shown in the chart where that row and column intersect.

	I	r_1	r_2
I	I	r_1	r_2
r_1	r_1	r_2	I
r_2	r_2	I	r_1

Since the set of rotational symmetries is finite, the table shows at a glance all composites, and the properties of composition can readily be determined. For example, the table shows that the composite of any two rotational symmetries of the triangle is a rotational symmetry of the triangle and that composition is an operation on this set.

We ask whether this set of rotational symmetries is a group under composition.

1) We agreed in a preceding section that composition of functions is *associative*.

2) There is an identity element in the set, the function I.

3) For each rotational symmetry, is there an inverse symmetry in the set? For example, given r_1, is there a symmetry such that the composite of r_1 and the symmetry is I? From the table we see that the inverse of r_1 is r_2 because

$$r_2 \circ r_1 = I \quad \text{and} \quad r_1 \circ r_2 = I.$$

144 functions in geometry: motions in the plane

In fact, for every symmetry there is an inverse. The inverse of r_2 is r_1 and the inverse of I is I.

Thus we find that under composition the rotational symmetries of an equilateral triangle have all the properties of a group. Furthermore, the table shows that composition is commutative in this group. We can then say that the group of the rotational symmetries of an equilateral triangle is a commutative group.

We can now ask whether the set of all the symmetries of an equilateral triangle is a group under composition. The table below shows the composition of any two symmetries. Verify some of its entries. For example, find the image of each vertex A, B, and C under $f_1 \circ r_2$ by first determining the image of a vertex under r_2 (rotate the triangle 240 degrees counterclockwise about X) and then determining the image under f_1 of this image (reflect it in line l_1, the line through vertex A). This gives the same image as that obtained under the mapping f_3; that is, $f_1 \circ r_2 = f_3$.

	I	r_1	r_2	f_1	f_2	f_3
I	I	r_1	r_2	f_1	f_2	f_3
r_1	r_1	r_2	I	f_2	f_3	f_1
r_2	r_2	I	r_1	(f_3)	f_1	f_2
f_1	f_1	f_3	f_2	I	r_2	r_1
f_2	f_2	f_1	f_3	r_1	I	r_2
f_3	f_3	f_2	f_1	r_2	r_1	I

We find that the symmetries of the equilateral triangle have all the properties of a group under composition. (The composite of any two of the symmetries is a symmetry; that is, composition is an operation on the set of symmetries of the triangle.)

1) Composition of symmetries of the triangle is associative.
2) There is an identity symmetry I.
3) For any symmetry there is an inverse symmetry.

Note that the group of all symmetries of the equilateral triangle is not commutative, since, for example, $f_1 \circ r_2$ is different from $r_2 \circ f_1$. We know that there are certain subsets of this set of symmetries which themselves form groups; one such *subgroup* is $\{I, r_1, r_2\}$, as we saw above. What are some other subsets that are groups? The group of all the symmetries of an equilateral triangle is itself a subgroup of the larger group of all rigid motions of the plane. In the group of symmetries of the equilateral triangle you will note that the number of elements in a subgroup is a factor of the number of elements in the group itself. This will be true in all finite groups.

There are other geometric figures that have more than one symmetry. For example, a square has four rotational and four reflective symmetries. In the exercises you will determine whether the set of symmetries of a square is a group under composition.

What at first seems to be idle speculation about an impractical mathematical notion turns out to be of prime usefulness in further studies and in research. For example, the study of groups of symmetries of space figures is closely related to theories of crystallography and molecular structure in chemistry and physics. It is also intrinsically interesting in its own right.

exercise set 4–4

1. Each of the following describes all the lines of the reflective symmetries of a particular quadrilateral. Describe the quadrilateral.
   a) Two diagonals
   b) One diagonal
   c) Two lines which are perpendicular bisectors of pairs of opposite sides
   d) Two lines which are perpendicular bisectors of pairs of opposite sides and two diagonals

2. Quadrilateral *PQRS* is a rectangle. Name its symmetries.
   a) Construct a table of the composites of its symmetries.
   b) Is composition of symmetries an operation?
   c) If so, name the properties of the operation and determine whether the structure is that of a group.

3. Quadrilateral *DEFG* is a rhombus. Name its symmetries.
   a) Construct a table of the composites of its symmetries.
   b) Is composition of symmetries an operation? If so, what are its properties?

4. $\{I, r_1, r_2\}$ is a subset of the symmetries of an equilateral triangle that is itself a group. What are some other subsets of the symmetries of an equilateral triangle that are groups under composition?

5. Show that in the group of symmetries of an equilateral triangle, the subset $\{I, f_1\}$ and the subset $\{I, f_2\}$ are subgroups. (Check that each group property is satisfied.)

6. In the group of symmetries of an equilateral triangle, solve the following equations, if possible. Example: $r_1 \circ x = f_2$, x is f_3.
   a) $r_1 \circ f_1 = x$
   b) $r_1 \circ x = f_1$
   c) $x \circ r_2 = f_3$
   d) $r_1 \circ f_1 = x \circ r_1$
   e) $f_3 \circ x = I$
   f) $f_1 \circ (f_2 \circ r_1) = x \circ r_1$

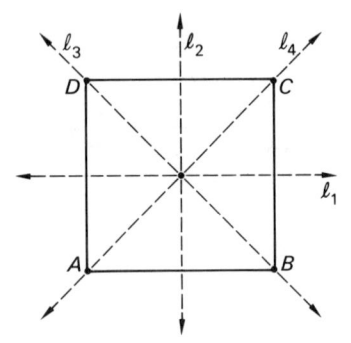

Figure 4–24

7. Consider the symmetries of a square (Fig. 4–24). Let I = identity function, r_1 = counterclockwise rotation $\frac{1}{4}$ of a revolution, r_2 = counterclockwise rotation $\frac{1}{2}$ of a revolution, r_3 = counterclockwise rotation $\frac{3}{4}$ of a revolution, f_1 = reflection in l_1, f_2 = reflection in l_2, f_3 = reflection in l_3, f_4 = reflection in l_4.
   a) Write each of the functions in this set of symmetries in the notation of the text. For example,
   $$f_2 = \begin{pmatrix} A & B & C & D \\ B & A & D & C \end{pmatrix}$$
   because
   $$f_2(A) = B, f_2(B) = A, f_2(C) = D, f_2(D) = C.$$
   b) Form a table of composites of the symmetries in this set. From the table, decide whether each symmetry has an inverse. Also decide whether the composition is commutative.

c) We see that the symmetries of a square form a group. Find all the subgroups of this group. Show that the number of elements in each subgroup is a divisor of 8 (the number of elements in the group).

d) Determine the elements $r_1, r_1 \circ r_1, r_1 \circ r_1 \circ r_1, \ldots$ (We can write these elements as $r_1, (r_1)^2, (r_1)^3, \ldots$) Which power of r_1 is the identity I? Show that all the powers of r_1 form a subgroup of the group of symmetries of the square.

8. In the group of symmetries of a square, solve the following equations, if possible.
   a) $f_3 \circ r_2 = x$
   b) $(f_2 \circ r_1) \circ x = I$
   c) $r_2 \circ x = f_3$
   d) $x \circ r_2 = f_3$
   e) $r_3 \circ x = f_2 \circ r_3$
   f) $x \circ r_3 = f_2 \circ r_3$
   g) $(x \circ x) \circ r_3 = f_4$
   h) $r_1 \circ (x \circ f_1) = r_2$

9. Show by a counterexample that composition of rigid motions is not commutative.

teaching questions and projects 4-3

1. Plan activities in which children describe the symmetries of familiar symbols such as N H E B Z O S T A.

2. Describe in detail a lesson in which children determine whether the set of symmetries of an equilateral triangle under composition of symmetries is an example of a group.

3. Discuss how you think the concept of a group should be introduced to children and at what level. How would you relate algebraic and geometric examples?

chapter five

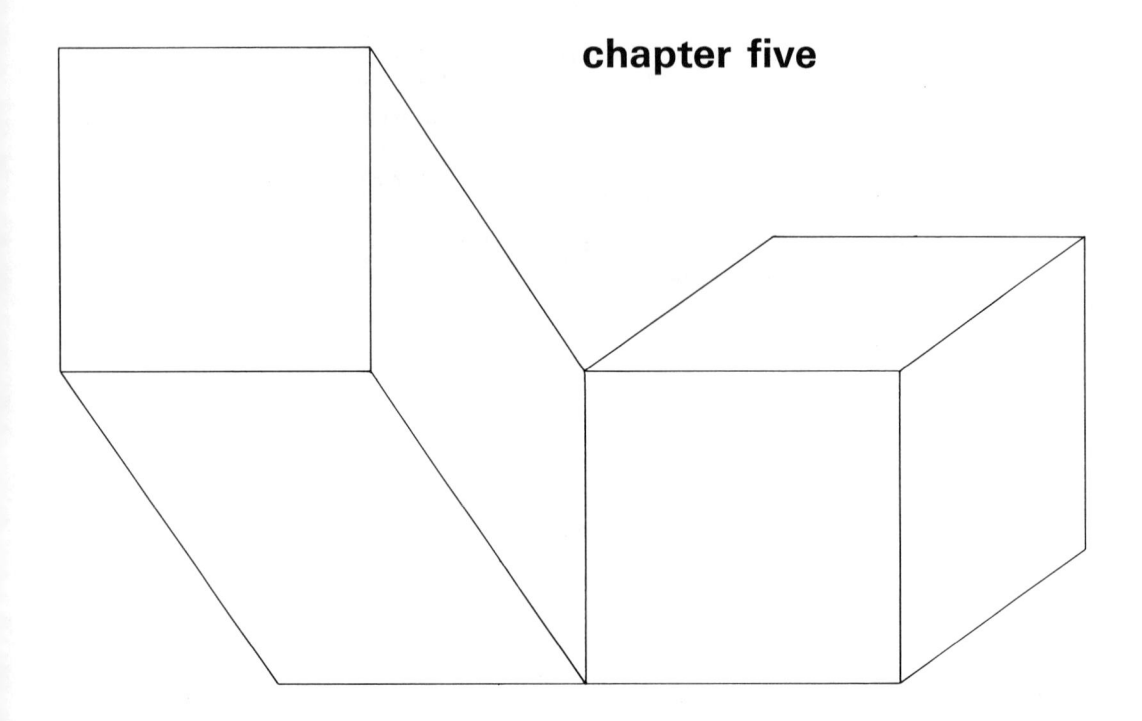

coordinates and vector geometry

5–1
the number line

introduction Through an elementary mathematics program there run two major streams: arithmetic and geometry. At the time that ideas such as points, lines, planes, distances between points, and intersections of lines are developed, there is also an extension of concepts of numbers to include the integers, the rational numbers and the real numbers, with the structure inherent in these number systems. Wherever possible, it is good mathematical practice to find and exploit connections between arithmetic and geometry.

Among the possible connections which can be employed effectively quite early in developing the ideas of mathematics is the number line, which is a simple device for assigning to the real numbers the points of a line. This device allows us to "visualize" sets of numbers as sets of points. A ruler is the commonest application of the number line; the points of the edge of the ruler mark the ends of line segments of various lengths, and the numbers associated with the points measure the line segments in inches, as in Fig. 5–1.

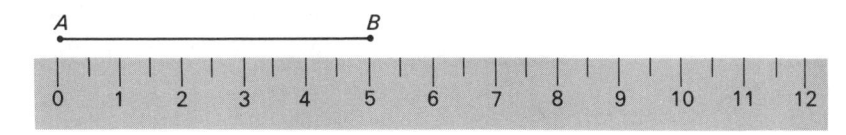

Figure 5–1

A is assigned to 0, B is assigned to 5; $\overline{AB}$ is 5 inches long

coordinate systems on the line There are many possible ways to assign to the real numbers the points of a line. Any assignment that is a one-to-one, onto function from the real numbers to the points of the line is called a *coordinate system on the line*. The number line is a particular coordinate system on the line which is the most natural function from numbers to points; the function assigns equispaced points to successive integers. We describe this function as follows:

Let the line be k, and select two distinct points O and U on k (Fig. 5–2).

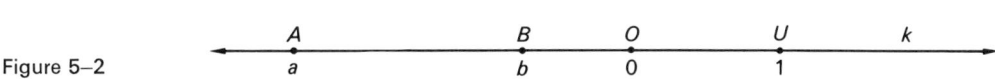

Figure 5–2

Let f be a one-to-one, onto function from the real numbers to the points of k such that:

1. $f(0) = O$ and $f(1) = U$. (That is, f maps 0 onto point O and f maps 1 onto point U.)
2. If $f(a) = A$ and $f(b) = B$, then line segment $\overline{AB}$ has measure $|b - a|$ with respect to the unit of measure $\overline{OU}$. (That is,

the segment from point A to point B is $|b - a|$* times as long as the unit segment $\overline{OU}$.)

3. If $a < b$ and $f(a) = A$, $f(b) = B$, then the ray $\overrightarrow{AB}$ has the same sense as the ray $\overrightarrow{OU}$. (That is, if $a < b$, then A is on the same side of B as O is of U.)

The above definition of the function f that we call the number line is an exercise in mathematical precision of language. The concept itself is so simple that it can be used quite easily and correctly by young children. It is less than simple only when we try to be precise in describing it. The following remarks reconcile the above definition with the simple notion it conveys.

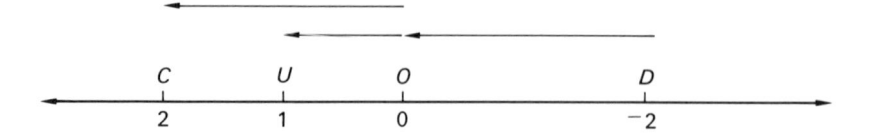

Figure 5–3

The unit point U is usually chosen to the right of the zero point O, as in Fig. 5–2, although this is merely a convention. The reader will find it interesting to choose U to the left of O in following the remainder of the remarks in this paragraph. Once the points O and U are assigned to the numbers 0 and 1, respectively, the segment $\overline{OU}$ acts as a unit of measure (see Fig. 5–3). Then the point, say C, which is assigned to another number, say 2, is located such that $\overline{OC} = 2\overline{OU}$ and for which $\overrightarrow{OC}$ has the same sense as $\overrightarrow{OU}$ (since $0 < 2$). The number $^-2$ is assigned to the point D, for which $\overline{OD} = |^-2| \overline{OU}$ and for which $\overrightarrow{DO}$ has the same sense as $\overrightarrow{OU}$ (since $^-2 < 0$). If we call the "right" of O the side of O on which U lies, then in Fig. 5–3 C is on the "right" of O and D is on the "left" of O. [*Caution:* If U is chosen to the left of O, as in Fig. 5–3, then "right" is left according to our definition. There is no inconsistency here except in the language. One might use language such as "above" and "below" O instead of

---

* For example, $|5 - 2| = |3| = 3$, $|2 - 5| = |^-3| = 3$. $|x|$ is called the absolute value of x, and is defined as: $|x| = x$ if $x \geq 0$; $|x| = ^-x$ if $x < 0$.

"right of" and "left of" O. Then C is above O and D is below O in Fig. 5–3.]

Note that the definition of function f relies on the notion of measurement of segments. In Postulate 2 (Chapter 3) we assumed that for every segment there is a positive real number measuring the segment with respect to a unit segment. Thus Part 2 of our definition of function f makes use of this measurement by letting $|b - a|$ be the measure of the segment $\overline{AB}$, where $f(a) = A$ and $f(b) = B$. Part 1 of our definition sets the unit of measure as $\overline{OU}$; part 3 requires that positive numbers be assigned points on the U side of O and negative numbers be assigned points on the other side of O.

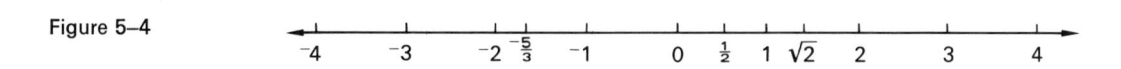

Figure 5–4

In practice we do not use the function notation. We simply assign points to numbers, as shown in Fig. 5–4. Thus every time we label a point of the number line with a numeral, we are assigning a point to a number according to the function f. Eventually this association between point and number becomes so intimate that we speak, for example, of the "$\frac{1}{2}$-point" or the "$^-3$-point."

Now it becomes clear why the number line is sometimes called the *equispaced coordinate system* on the line. Distances between pairs of consecutive integer points are equal.

The function f is from the *real* numbers to points of the line. So far we have described the locations only of integer points. The locations of rational points and irrational points also follow from the definition of f. For example, what point D does f assign to $^-\frac{5}{3}$? Since $^-\frac{5}{3} < 0$, the $^-\frac{5}{3}$-point must be "below" the 0-point on the line. Also the segment $\overline{OD}$ must be $|^-\frac{5}{3}|$ as long as $\overline{OU}$. We trisect the segment $\overline{OU}$ and form a segment $\overline{OD}$ that is 5 times the length of $\frac{1}{3}\overline{OU}$ so that $\overrightarrow{OD}$ has sense opposite to that of $\overrightarrow{OU}$. Clearly each rational number $\frac{p}{q}$, for p any integer and q any counting number, will be mapped onto a point of the line k by this procedure. The irrational numbers are mapped onto points of line k in the manner of the discussion of Section 2–9. (The reader might profit by rereading Sections 2–8 and 2–9.)

graphs on the number line Recall that the number line is a coordinate system on the line. We say that if f maps the number a onto point A, then a is the f-coordinate of A. Not only does the number line help us to visualize the natural order of the real numbers, but it also provides a way to single out special subsets of the real numbers. For example, the sentence

$$3x - 6 < x, \quad x \text{ an integer,}$$

has the solution set $\{x \text{ an integer} \mid x < 3\}$,* which can be determined by trial and error or by more formal means. This set can be graphed on the number line; that is, the points of the line corresponding to the numbers in the set can be shaded or singled out in some way. The set of shaded points in Fig. 5–5 is called the *graph* of the set of numbers. (Note that only a part of the figure can be drawn; the three dots on the left indicate that the graph continues to the left without end.)

Fig. 5–5. Graph of the set $\{x \text{ an integer} \mid x < 3\}$.

The sentence

$$3x - 6 < x, \quad x \text{ a real number,}$$

has the solution set $\{x \text{ real} \mid x < 3\}$, and its graph is shown in Fig. 5–6. This graph consists of all the points to the left of the 3-point. Again only a portion of the graph can be shown on the figure.

Fig. 5–6. Graph of the set $\{x \text{ real} \mid x < 3\}$.

---

* This notation is read "the set of all integers x such that x is less than 3."

Figure 5-7

translations of the line The number line is also useful in representing translations of the line. We have described a translation as a function that maps the plane onto the plane in such a way that line segments map onto congruent and parallel segments. Let us restrict the domain and range of the translation to the points of the number line. Then if the translation t maps point A onto point B (Fig. 5–7), it will map each point X of the line onto a point Y such that $XY = AB$ and Y is on the same side of X as B is of A. Here the utility of the number-line concept becomes evident. Since each point of the number line has a real-number coordinate, we can describe the translation t in terms of coordinates (that is, real numbers) as follows:

Let the coordinates of points A, B, X, Y be the numbers a, b, x, y, respectively (see Fig. 5–8). Now we must distinguish between the *length* of $\overline{AB}$ (that is, the distance *between* A and B) and the distance *from* A *to* B. The length of $\overline{AB}$ is the positive number $|b - a|$. The distance from A to B is the number $b - a$, which is positive if B is above A, negative if B is below A. [*Note:* Distance *between* points is the distance described in Postulate 2, whereas distance *from* one point *to* another is a directed distance.]

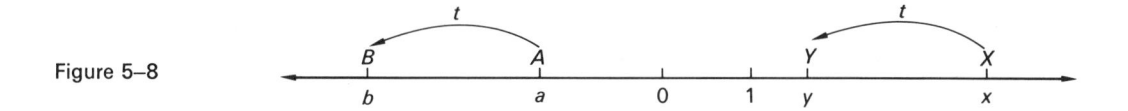

Figure 5-8

Then the distance from A to B is the number $b - a$ and the distance from X to Y is $y - x$. Let c be the real number $b - a$. (In the case shown in Fig. 5–8, c is a negative number.) Since the translation t preserves the lengths and directions of line segments, we see that the distance from X to Y must also be c, so that

$$y - x = c;$$

that is,

$$y = x + c.$$

This tells us that if translation t takes point A (with coordinate a) to point B (with coordinate b), then t takes any point X (with coordinate x) to the point Y with coordinate $x + c$, where

$$c = b - a.$$

In this way a translation t of the line can be described in terms of coordinates of the points of the line. The notation

$$x \xrightarrow{t} x + c$$

indicates that t takes any point with coordinate x to the point with coordinate $x + c$. Thus the translation t "marches each point" $|c|$ units to the right if $c > 0$; t "marches each point" $|c|$ units to the left if $c < 0$. (We shall hereafter use the number line with O and U placed so that left and right have their usual meanings. Note that $|c|$ is either a positive number or zero.)

For example, let the translation t map the 3-point onto the 7-point. Then $c = 4$, since $7 - 3 = 4$, and $3 \xrightarrow{t} 3 + 4$. For each point with coordinate n,

$$n \xrightarrow{t} n + 4,$$

so that, for example, $0 \xrightarrow{t} 4$, $5 \xrightarrow{t} 9$, $^-4 \xrightarrow{t} 0$, $^-2 \xrightarrow{t} 2$, and $^-\tfrac{4}{3} \xrightarrow{t} \tfrac{8}{3}$. We see that this translation t is completely determined by the number 4, because each point is moved to the right $|4|$ units. Thus we can call it the 4-translation and write $n \xrightarrow{4} n + 4$. This translation can be shown on the number line with any arrow $|4|$ units long pointing right. The tail of this arrow is at some point A of the line and its head is at $t(A)$. In Fig. 5–9, A is the $^-2$-point and $t(A)$ is the 2-point.

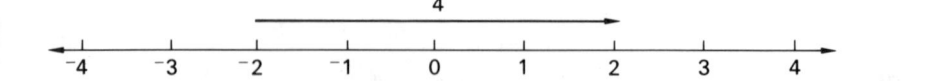

Figure 5–9

Consider another example of this kind. Let a translation map the 6-point onto the $\tfrac{9}{2}$-point. Then $c = ^-\tfrac{3}{2}$ and $6 \to 6 + ^-\tfrac{3}{2}$; we identify this translation with the number $^-\tfrac{3}{2}$, and for each point with coordinate n,

$$n \xrightarrow{^-\tfrac{3}{2}} n + ^-\tfrac{3}{2},$$

so that $0 \to ^-\tfrac{3}{2}$, $\tfrac{7}{2} \to 2$, $\tfrac{3}{2} \to 0$, $^-\tfrac{5}{3} \to ^-\tfrac{19}{6}$, ... This trans-

Figure 5–10

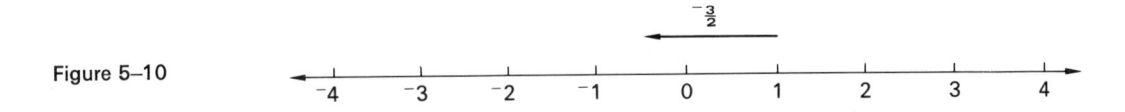

lation can be shown with any arrow $|{}^-\tfrac{3}{2}|$ units long pointing left (Fig. 5–10).

Corresponding to each real number there is a translation of the line, and, conversely, for each translation of the line there is a real number representing it.

> For any real number c there is a translation of the line, $n \xrightarrow{c} n + c$, which maps each point with coordinate n onto the point with coordinate $n + c$.

composition of translations We know that the composite of two translations is a translation. For example, consider the composite of the two translations of the line corresponding to the numbers 4 and ⁻6:

$$n \xrightarrow{4} n + 4; \qquad n \xrightarrow{-6} n + {}^-6.$$

Call the composite t. Then the 4-translation maps n onto $n + 4$ and the ⁻6-translation maps $(n + 4)$ onto $(n + 4) + {}^-6$, so that

$$n \xrightarrow{t} (n + 4) + {}^-6 \quad \text{or} \quad n \xrightarrow{t} n + (4 + {}^-6).$$

That is, the composite of the translations 4 and ⁻6 is the translation $(4 + {}^-6)$, or the translation ⁻2.

Graphically this can be shown by a movement of the line $|4|$ units to the right followed by a movement $|{}^-6|$ units to the left. The result is a movement of the line $|4 + {}^-6|$ units to the left.

Since a translation of the line is determined when we know the image of just one point, we can concentrate on the 0-point. The 4-translation maps 0 onto 4 and then the ⁻6-translation maps 4 onto ⁻2. (See either Fig. 5–11 or 5–12.) Hence the composite translation maps 0 onto ⁻2, and we find that the com-

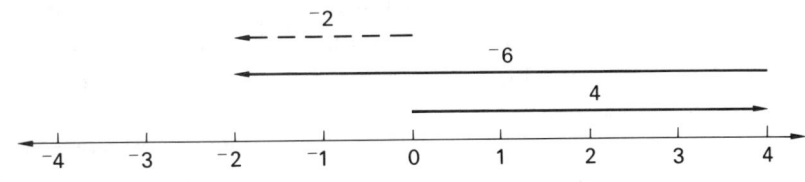

Figure 5–11

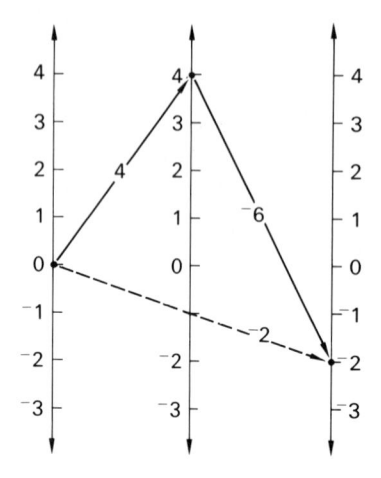

Figure 5-12

posite is the ⁻2-translation. In general:

> The composite of the a-translation and the b-translation is the $(a + b)$-translation.

remarks about addition of integers There are various ways to introduce the addition of integers at the elementary school level.

1. Integers can be associated intuitively with gains and losses in games, and sums are then associated with accumulated gains and losses. For example, a profit of 10 dollars in the morning followed by a loss of 17 dollars in the afternoon results in a net loss of 7 dollars for the day; this suggests that $10 + {}^{-}17 = {}^{-}7$.

2. Integers can be associated intuitively with movements on the number line. Indeed, this is probably the most direct and natural way to learn addition of integers. Integers may be interpreted as "marching orders" on the number line, starting at the 0-point, and ending at a point which suggests the sum of the integers. On the other hand, when we consider the geometric problem of composing two translations of a line, the problem can be coordinatized and reduced to a problem in adding numbers. Here we have a good example of the blending of arithmetic and geometry in which, in varying contexts, one can be of service to the other.

3. The idea that each integer corresponds to a translation of the line will make the early intuitions clear, and compositions of translations will yield all the desired properties of addition of integers. Recall that addition of integers has all the properties of a commutative group. These properties are:

a) Composition of translations is associative and commutative; thus the corresponding addition of integers is associative and commutative.

b) The identity mapping corresponds to the integer 0; this implies that for any integer a,
$$a + 0 = 0 + a = a.$$

c) For each translation t of the line there is an inverse translation t' such that the composite of t and t' is the identity mapping. Thus for each integer c there is an integer ^-c (the opposite of c or the additive inverse of c) such that $c + (^-c) = 0$.

Of course, the above properties hold for addition of all real numbers, since each translation of the line corresponds to a real number, and conversely.

exercise set 5–1

1. Given the number line (Fig. 5–13) with points marked for the integers.

   Figure 5–13

   a) Construct the point for $\frac{5}{2}$, for $^-\frac{7}{3}$, for $\sqrt{2}$, for $^-\sqrt{3}$. For example, to construct a segment $\overline{AC}$ that is $\frac{1}{3}$ as long as a segment $\overline{AB}$, proceed as shown in Fig. 5–14. (On any other line l through A and with any unit segment $\overline{AD}$, determine a segment $\overline{AE}$ three times the length of the $\overline{AD}$. Then construct $\overline{DC}$ parallel to $\overline{BE}$. Then $\overline{AC}$ is $\frac{1}{3}$ the length of $\overline{AB}$.)

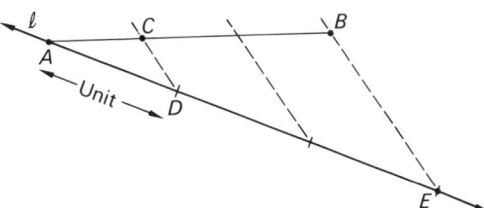

   Figure 5–14

   b) Given the points for $\frac{1}{3}$ and $\frac{1}{4}$, explain how to find the points for $\frac{1}{6}$, $\frac{1}{8}$, and $\frac{1}{12}$.

2. Determine:
   a) $|-\frac{3}{2} + \frac{3}{4}|$
   b) $|3(^-2 + 3)|$
   c) $|^-4| + |3|$
   d) $|^-4 + 3|$
   e) $|a|$ if $a < 0$
   f) $|^-3 \times 4|$
   g) $|^-3| \times |^-4|$

3. Show that $|a - b| = |b - a|$.

4. For each of the following sentences, determine its solution set and exhibit its graph on the number line.

   a) $3x + 5 = 2$, x an integer
   b) $x + 2 < 8$, x a whole number
   c) $x + 2 < 8$, x an integer
   d) $x + 2 < 8$, x a real number
   e) $2 + 3x > 4$, x a real number
   f) $3 + x < 3x - 5$, x a real number

5. Each of the following is a translation of the line. Determine its effect on any point with coordinate n.

   Example: t is a translation such that $t(^-3) = 2$. (This is another notation for $^-3 \xrightarrow{t} 2$.) Then, since $^-3 + 5 = 2$, we have $t(n) = n + 5$, and t is the 5-translation.
   a) v is a translation such that $v(^-2) = ^-6$.
   b) r is a translation such that $\frac{3}{2} \xrightarrow{r} -\frac{1}{2}$.
   c) w is a translation such that $w(-\frac{3}{4}) = ^-2$.
   d) z is a translation such that $3 \xrightarrow{z} 3 - \sqrt{2}$.

6. Let points A and B on the number line have coordinates 3 and $-\frac{5}{3}$, respectively.
   a) What is the distance between A and B, that is, the length of $\overline{AB}$?
   b) What is the directed distance from A to B?
   c) What is the directed distance from B to A?
   d) What is the coordinate of the midpoint of $\overline{AB}$?
   e) Given that the coordinate of A is a and the coordinate of B is b; show that the coordinate m of the midpoint of $\overline{AB}$ is $\frac{1}{2}(a + b)$.

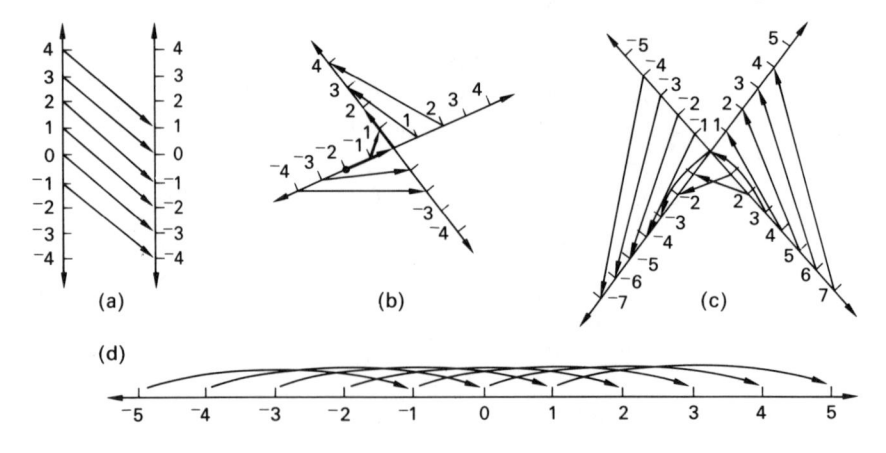

Figure 5–15

7. Figure 5–15 shows some diagrams that suggest translations of the line. In each case, identify the translation.

8. Determine the composite of the translations t and s, given that:
   a) $t(^-2) = 6$, $s(6) = ^-5$
   b) $t(0) = ^-7$, $s(^-7) = 3$
   c) $t(^-2) = 1$, $s(^-2) = ^-3$
   d) $t(^-\frac{1}{2}) = \frac{3}{4}$, $s(3) = ^-\frac{1}{2}$
   e)

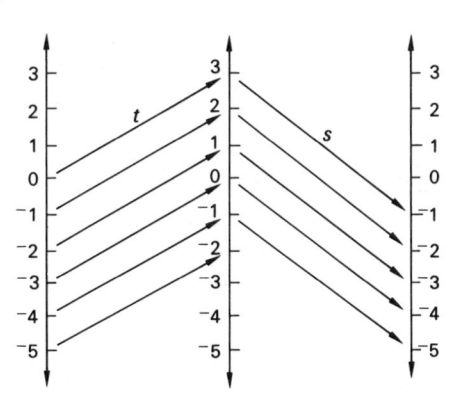

Figure 5–16

9. Explain how the composition of translations of a line is related to addition of real numbers. Illustrate with arrow diagrams on a line.

5–2
the number plane

coordinate systems on the plane First experiences with the assignment of points of the plane to ordered pairs of numbers, relative to a pair of intersecting number lines, are not necessarily of a formal nature. Simple games, such as coordinate tic-tac-toe, result in an intuitive idea of *the number plane*.

On a more mature level we treat *the number plane* as follows. On a plane draw two intersecting lines k_1 and k_2 (see Fig. 5–17). On each line construct a coordinate system with its 0-point at the point of intersection of the lines. Then define a one-to-one onto mapping from the set of all ordered pairs (a, b) of real numbers to the points of the plane as follows: Let the pair of numbers (a, b) be mapped onto a point P by locating the point on line k_1 whose coordinate is a and the point on line k_2 whose coordinate is b, and then finding the point P which is the intersection of the line through the a-point parallel to k_2 with the line through the b-point parallel to k_1.

160 coordinates and vector geometry

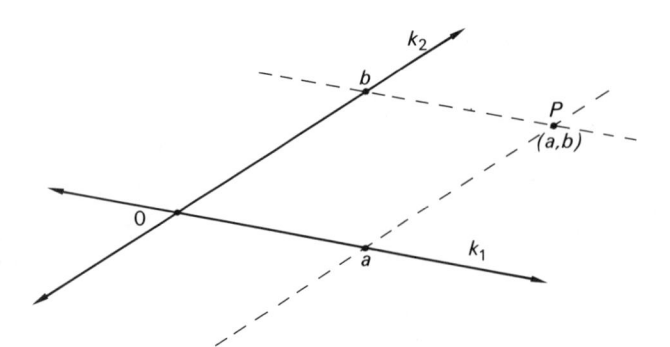

Figure 5–17

This construction defines a function that is one-to-one and onto, since each pair of lines parallel to k_2 and k_1, respectively, must intersect in exactly one point of the plane, and each point of the plane determines exactly one pair of lines parallel to k_2 and k_1, respectively. Such a function is called a *coordinate system on the plane*. If the coordinate systems on lines k_1 and k_2 are equispaced (that is, if they are each copies of the number line) with the same unit of measure, and if k_1 and k_2 are perpendicular, the resulting coordinate system on the plane is called a *number plane* or a *Cartesian coordinate system* (Fig. 5–18). The lines k_1, k_2 are called the *axes* of the coordinate system.

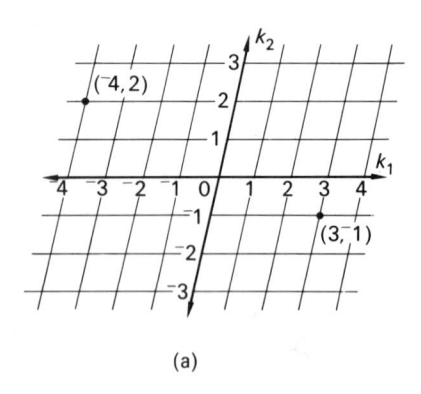

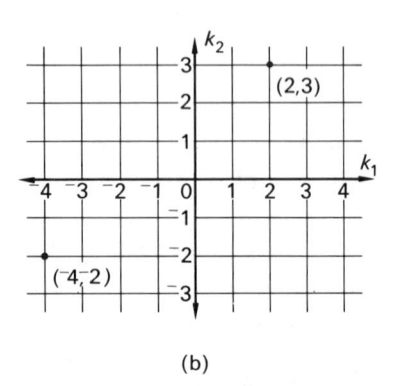

Fig. 5–18. (a) A coordinate system on the plane. (b) The number plane.

Note that this definition of the number plane allows two different coordinate systems and hence cannot strictly be called "the" number plane. One of these, shown in Fig. 5–18(b), is called a right-handed coordinate system: A counterclockwise rotation about the (0, 0) point through 90° takes the 1-point of k_1

to the 1-point of k_2. A left-handed system is one in which a clockwise 90° rotation about (0,0) takes the 1-point of k_1 to the 1-point of k_2. When we refer to "the" number plane we mean the right-handed number plane, just as "the" number line refers to the right-handed system on the line in which the 1-point is on the right side of the 0-point.

Points of the plane and ordered pairs of real numbers become intimately associated throughout mathematics in many contexts. Some examples follow.

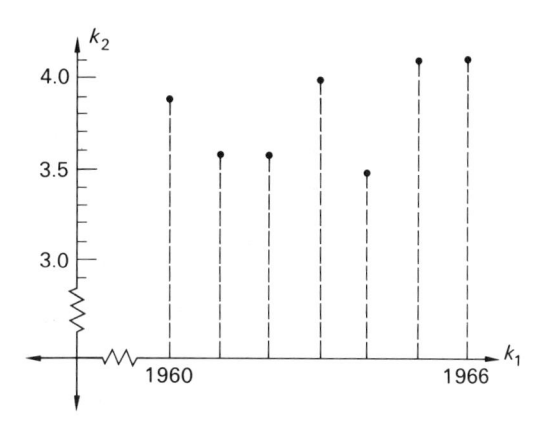

Figure 5–19

graphing of information The usual problems of graphing information are simplified if the information is assembled as a set of ordered pairs. For example, the graph of the yearly production of corn in the United States from 1960 to 1966 is the set of points in a coordinate system corresponding to the set of pairs,

$$\{(1960, 3.9), (1961, 3.6), (1962, 3.6), (1963, 4.0),$$
$$(1964, 3.5), (1965, 4.1), (1966, 4.1)\},$$

where the first entry in each pair is the year and the second entry is the number of billions of bushels of corn produced. The relative heights of the points in the graph is sometimes dramatized by such methods as erecting bars to the points (bar graph) or by drawing stalks of corn with their tops at the points (picture graph). The units on the axes are chosen in any convenient way; in Fig. 5–19 a coordinate system is chosen with the unit on k_1 larger than the unit on k_2. In any of these schemes the essential idea is the one-to-one correspondence between the ordered pairs and the points.

graphs of truth sets In various situations we use mathematical sentences in which two numbers are unknown, such as

$$x + 2y = 6, \quad y = x + 2, \quad y > x + 1, \quad xy = 12,$$

where x and y are elements of specific sets of numbers. For convenience, let us call these sentences "two-variable sentences." For example, if the perimeter of a rectangle is 12 inches and the lengths of the sides are x inches and y inches, then the sentence

$$2x + 2y = 12, \quad x \text{ and } y \text{ positive real numbers,}$$

describes the dimensions of all such rectangles. We note that *pairs* of numbers are needed to make this sentence true, one number for x and one for y. Let us agree to write such pairs always with the number for x first, so that we need *ordered pairs* of numbers to make the mathematical sentence true. One such ordered pair is $(1, 5)$, since $2 \cdot 1 + 2 \cdot 5 = 12$ is a true sentence, whereas $(6, 1)$ does not make the sentence true. ($2 \cdot 6 + 2 \cdot 1 = 12$ is a false sentence.)

The set of all the ordered pairs that make a two-variable sentence true is called its *truth set* (or *solution set*). For example, the sentence

$$x + 2y = 6, \quad x \text{ and } y \text{ whole numbers}$$

has truth set

$$\{(0, 3), (2, 2), (4, 1), (6, 0)\}.$$

Note that for this sentence the numbers for x and y are restricted to the set of whole numbers. The sentence

$$x + 2y = 6, \quad x \text{ and } y \text{ integers}$$

has truth set

$$\{\ldots, (^-4, 5), (^-2, 4), (0, 3), (4, 1), (6, 0), (8, ^-1), \ldots\}.$$

The sentence

$$x + 2y = 6, \quad x \text{ and } y \text{ real numbers}$$

has a truth set whose members cannot all be enumerated. We can describe the truth set as

$$\{(x, y), x \text{ and } y \text{ real} \mid x + 2y = 6\},$$

which is read "the set of all ordered pairs (x, y) of real numbers such that $x + 2y = 6$ is true." Note that this is merely a restatement of the definition of the truth set. But since each member of the truth set is an ordered pair, we can determine the set of all points in the number plane corresponding to the members of the truth set. That is, we can determine the graph of the sentence.

The set of points of the number plane corresponding to the members of the truth set of a two-variable sentence is called the *graph* of the sentence. Graphs of the sentence $x + 2y = 6$ for the three different sets of numbers discussed above are shown in Fig. 5–20.

Fig. 5–20. (a) Graph of $x + 2y = 6$, x and y whole numbers (all of graph shown). (b) Graph of $x + 2y = 6$, x and y integers (part of graph shown). (c) Graph of $x + 2y = 6$, x and y real numbers (part of graph shown).

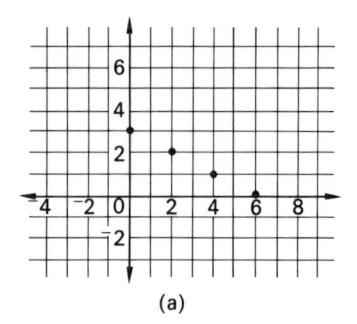

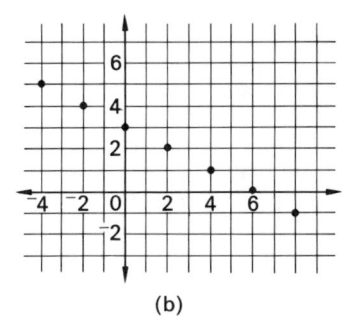

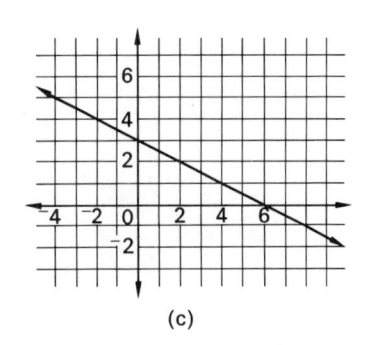

Note that each of the graphs in Fig. 5–20 seem to have their points on a line, although we cannot show all points of graphs (b) and (c) because they continue without end.

Consider the two-variable sentence

$$y > x + 1, \quad x \text{ and } y \text{ whole numbers.}$$

Part of the graph of this sentence is shown in Fig. 5–21. For example, (5, 6) is not in the graph, since "$6 > 5 + 1$" is a false sentence, whereas (15, 17) is in the graph, since "$17 > 15 + 1$" is true.

164 coordinates and vector geometry

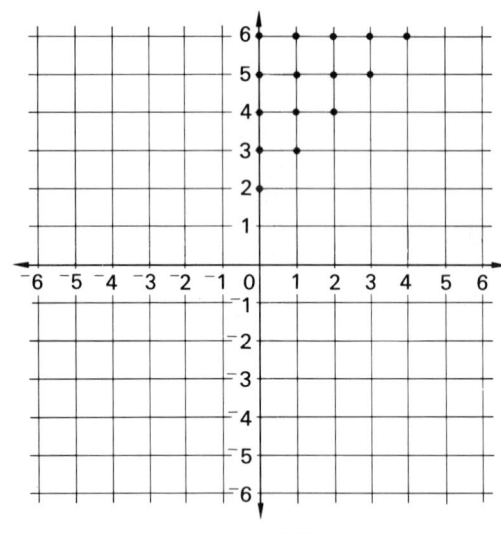

Fig. 5–21. Graph of $y > x + 1$, x and y whole numbers (part of graph shown).

The sentence

$$y > x + 1, \qquad x \text{ and } y \text{ real numbers}$$

consists of all points in a half-plane, partly shown by the shaded portion in Fig. 5–22.

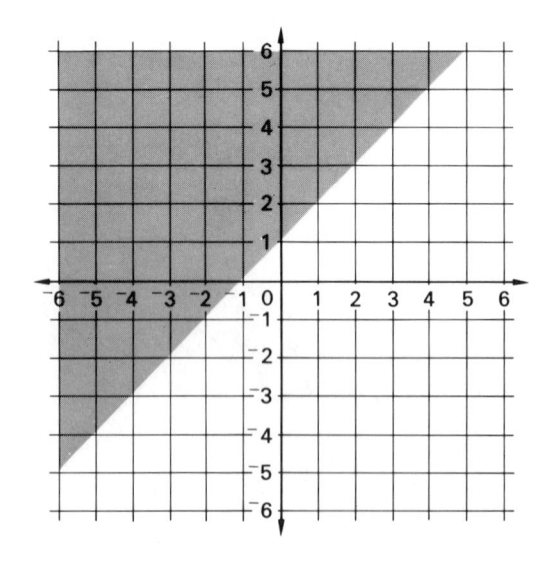

Fig. 5–22. Graph of $y > x + 1$, x and y real numbers (part of graph shown).

examples of sentences and graphs We can draw graphs of truth sets as soon as we learn to associate points of the plane with ordered pairs of numbers. For example, when listing pairs of whole-number addends of 10, we are really finding the truth set of the sentence

(1) $x + y = 10,$ x and y whole numbers.

This sentence provides the experience of "seeing" the pairs of addends as points of a graph (see Fig. 5–23a).

Also, when listing pairs of whole-number factors of 12, we are really finding the truth set of the sentence,

(2) $xy = 12,$ x and y whole numbers.

Graphs of sentences (1) and (2) are shown in Fig. 5–23.

Fig. 5–23. (a) Graph of $x + y = 10$, x and y whole numbers. (b) Graph of $xy = 12$, x and y whole numbers.

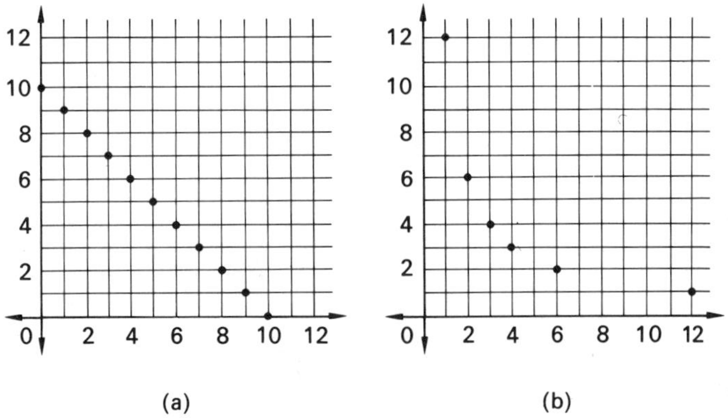

Various other common situations suggest two-variable sentences, and graphs of these sentences help us to analyze the situations. For example:

Buying potatoes at 13 cents a pound, with a 2-cent charge for the bag, suggests the sentence

(3) $y = 13x + 2,$ x and y positive integers,

where y represents the number of cents in the cost and x the number of pounds of potatoes.

Finding areas of squares suggests the sentence

(4) $y = x^2,$ x and y positive real numbers,

where x represents the measure of a side of the square and y the number of square units in the area.

Driving at 10 miles per hour suggests the sentence

(5) $\quad y = 10x, \quad$ x and y positive real numbers,

where x represents the number of hours and y the number of miles.

Driving a distance of 10 miles suggests the sentence

(6) $\quad xy = 10, \quad$ x and y positive real numbers,

where x represents the number of miles per hour and y the number of hours.

Sketch the graphs for sentences (3) through (6), and note that the graphs of (1), (3), and (5) appear to lie on lines, whereas the graphs of (2), (4), and (6) do not. Graphs of (1), (3), and (5) are called *linear graphs*. Since we shall make this distinction among graphs, we need to look for patterns of sentences that yield linear graphs. Examine sentences (1), (3), and (5).

$$x + y = 10, \qquad x \text{ and } y \text{ whole numbers,}$$
$$y = 13x + 2, \qquad x \text{ and } y \text{ positive integers,}$$
$$y = 10x, \qquad x \text{ and } y \text{ positive real numbers.}$$

First we decide from examination of such sentences that in order to have a linear graph a sentence must be an equation, and next that the equation can be written in a certain form, which we express here generally as

$$ax + by = c,$$

where a, b, and c are fixed real numbers such that a and b are not both zero.

Whether or not an elementary school mathematics program should go further than this in the analytic geometry of lines in a plane (through analysis of sentences with linear graphs) depends on various factors. Certainly there are opportunities at all levels for discovery without formal development. For example, children can discover what sentences have parallel linear graphs or perpendicular linear graphs and how to predict this parallelism or perpendicularity; they can find a way to determine where two lines intersect or where they cross the axes; they can learn to describe half-planes in terms of sentences involving inequalities. The reader is invited to review (or rediscover) some of these results in the geometry of lines from the viewpoint of the equations of the lines.

coordinate systems in space The extension of the concept of coordinate systems on lines and planes to coordinate systems in space is a natural one. Recall that a coordinate system on a line is a one-to-one, onto function from the real numbers to the points of the line, and that a coordinate system on a plane is a one-to-one, onto function from the set of all ordered pairs of real numbers to the points of the plane. We follow the pattern and define a coordinate system in space as a one-to-one, onto function from the set of all ordered *triples* of real numbers to the points of space.

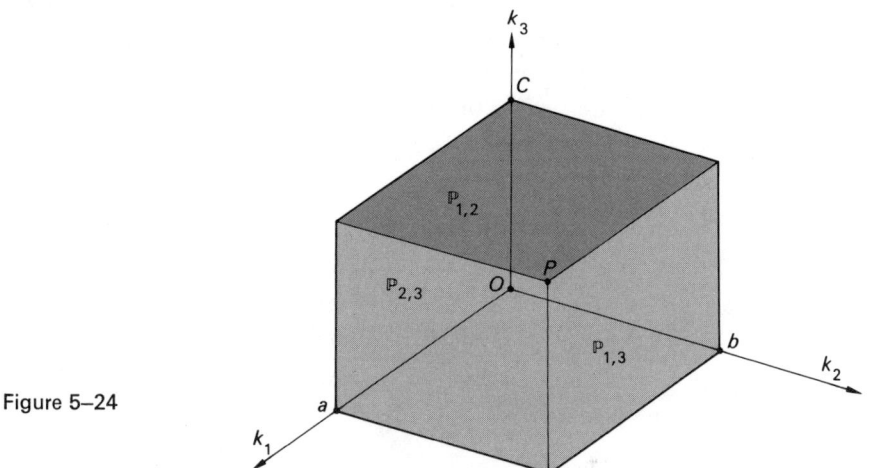

Figure 5–24

Choose any three lines k_1, k_2, k_3 intersecting in a point O such that the lines are not all in one plane (Fig. 5–24). On each line set up a coordinate system with its 0-point at O. Then to each ordered triple of real numbers (a, b, c) we can assign a point P in space as follows: Find the point on k_1 with coordinate a, the point on k_2 with coordinate b, and the point on k_3 with coordinate c. Through the a-point there is one plane $\mathbb{P}_{2,3}$ parallel to the plane of k_2 and k_3; through the b-point there is one plane $\mathbb{P}_{1,3}$ parallel to the plane of k_1 and k_3; through the c-point there is one plane $\mathbb{P}_{1,2}$ parallel to the plane of k_1 and k_2. These three planes intersect in exactly one point P (why?), and we assign P to the ordered triple (a, b, c). Conversely, given any point P in space, there is exactly one set of three planes through P parallel to the planes of k_2 and k_3, k_3 and k_1, and k_1 and k_2, respectively, each of which intersects an axis in one point, giving an ordered triple of coordinates of points on k_1, k_2, k_3.

168 coordinates and vector geometry

If the axes k_1, k_2, k_3 are perpendicular in pairs such as in Fig. 5–24, and if the units on the axes are the same length, we call the resulting coordinate system *a number space*.

At this point we could profitably continue the extension by considering sentences with three unknowns, such as

$$2x + y + z = 4, \quad x, y, z \text{ real numbers,}$$

and their truth sets, in this case a set of ordered triples of real numbers including such triples as (1, 1, 1) and (⁻2, ⁻1, 9). The graph of such a sentence is a set of points in space; in the above example, the set of points is a plane. The reader is invited to draw this graph, where the variables are the ordered triple x, y, z, and to extend some other concepts of graphs in a number plane to graphs in a number space.

exercise set 5–2

1. For each of the following coordinate systems on the plane, find the point for each of the ordered pairs (⁻2, 1), (1, ⁻2), (2, ⁻1), (⁻1, 2), (3, 0), (0, ⁻3).

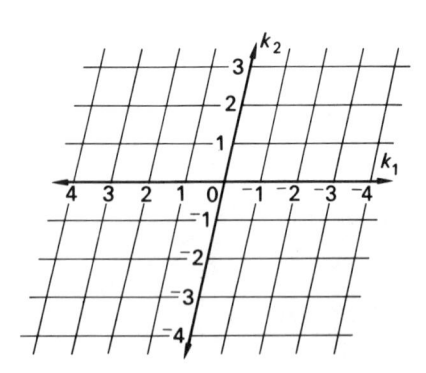

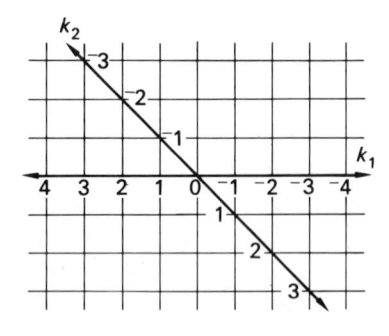

Figure 5–25

2. a) Drop two dice 36 times, and keep a record of the number of outcomes for each sum. List this information as a set of ordered pairs [e.g., the ordered pair (10, 3) means that the sum 10 occurred 3 times] and draw a graph of this information on the number plane.
   b) Drop four coins 32 times, record the number of times you get 0, 1, 2, 3, or 4 heads, and draw a graph of this information.
   c) Describe the first-class postage function as a set of ordered pairs, where the first entry in a pair is a positive real number giving

the weight in ounces and the second is a positive integer giving the cost of the postage in cents. Draw a graph of this function (that is, a graph of the set of ordered pairs) on a coordinate system with convenient units on the axes.

3. For each of the following sentences, find some ordered pairs in its truth set and then draw the graph of the sentence in the number plane.
   a) $3x + y = 6$, x and y whole numbers
   b) $3x + y = 6$, x and y integers
   c) $3x + y = 6$, x and y real numbers
   d) $y = 2x + 3$, x and y real numbers
   e) $y = x \cdot x$, x and y integers
   f) $y < 3 - x$, x and y whole numbers
   g) $x + y < 3$, x and y real numbers
   h) $xy = 24$, x and y integers
   i) $xy < 24$, x and y counting numbers
   j) $2x + 3y = 6$, x and y real numbers

4. For each of the following truth sets or graphs find a corresponding two-variable sentence.
   a) $\{(0, 4), (1, 3), (2, 2), (3, 1), (4, 0)\}$
   b) $\{(1, 6), (2, 3), (3, 2), (6, 1)\}$
   c) $\{(0, 3), (0, 2), (0, 1), (0, 0), (1, 2), (1, 1), (1, 0), (2, 1), (2, 0), (3, 0)\}$
   d)   e)   *f)

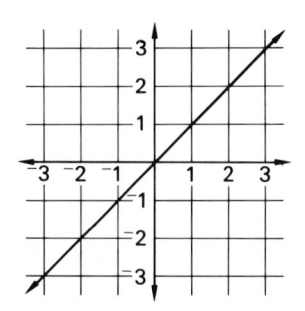

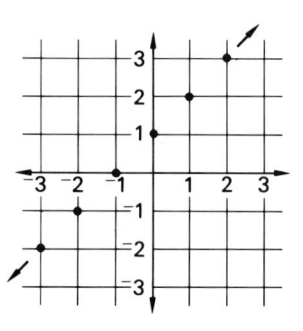

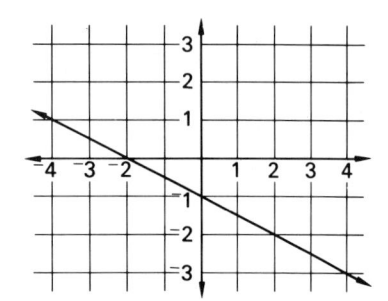

Figure 5–26

5. Let f be a function from the real numbers to the real numbers such that for any number a,
$$a \xrightarrow{f} a - 3$$
Describe f as the set of ordered pairs $(a, a - 3)$ for any real number a. [For example, $(3, 0)$, $(4, 1)$ and $(\frac{7}{2}, \frac{1}{2})$ are in this set.] Draw a graph

of this set of ordered pairs. In other words, draw a graph of the function f. [Note that this is also the graph of the sentence

$$y = x - 3, \quad x \text{ and } y \text{ real numbers.}]$$

6. a) Draw a graph of the function g, where

$$a \xrightarrow{g} 2 - 3a, \quad \text{for any real number } a.$$

b) What two-variable sentence has the same graph? [*Hint:* Given that (x, y) is any ordered pair in the truth set of the sentence, then we must have $x = a$ and $y = 2 - 3a$.]

7. Draw a graph of each of sentences (3), (4), (5), and (6), on pages 165 and 166, and verify that certain of these graphs lie on lines.

8. Consider the two-variable sentence

$$ax + by = c, \quad x \text{ and } y \text{ real numbers,}$$

where a, b, c are fixed real numbers such that a and b are not both 0.
a) For each of the following specified values of a, b, c, show that the resulting sentence has a linear graph.
   i) $a = 0, b = 1, c = 2$ \qquad ii) $a = 1, b = 0, c = 2$
   iii) $a = 1, b = {}^-2, c = 0$ \qquad iv) $a = 1, b = 0, c = 0$
   v) $a = 0, b = 1, c = 0$
b) Discuss the sentence $ax + by = c$ when $a = 0$ and $b = 0$. What is its truth set? Is its graph linear?

9. Draw graphs of each pair of sentences on one number plane. Are the graphs for each pair parallel? perpendicular?
a) $2x + y = 4, x + 2y = 4$
b) $2x + y = 4, 2x + y = 2$
c) $2x + y = 4, x - 2y = 3$
d) $2x + y = 4, 4x + 2y = 5$

10. On the basis of your responses to Exercise 9, make conjectures concerning the forms of sentences with parallel graphs; with perpendicular graphs. Try other examples to test your conjectures.

11. Draw graphs of these two sentences, and from the graphs attempt to find an ordered pair common to the solution sets of the two sentences:

$$2x + y = 5; \quad x - 3y = 6.$$

12. Write a two-variable sentence for each of the following situations, with a description of each of the variables in the sentence.
a) Buying syrup at 8 cents an ounce, with a 25-cent charge for the bottle.

b) Determining the dimensions of any rectangle with area of 24 square feet.
c) Determining the volume of rectangular prisms which have height 4 inches and width 3 inches.
d) Driving an automobile at constant speed for 5 hours.
e) Driving an automobile at constant speed a distance of 20 miles.

13. Give several physical interpretations of each of these sentences (i.e., reverse the process of Exercise 12).
a) $y = 16x$ b) $y = 6x + 5$ c) $xy = 18$ d) $y = 2x - 3$

5–3
vectors in the plane

translations; vectors; ordered pairs of numbers Among the rigid motions of the plane described in Chapter 4, the translations are most easily represented in terms of coordinate systems. We have noted that each translation of the line is determined by a real number. We shall see that each translation of the plane is determined by an ordered pair of real numbers.

Let t be a translation of the plane that maps point A onto point B (Fig. 5–27). Then for any point C, translation t maps C onto a point D such that $AB = CD$, $\overline{AB} \parallel \overline{CD}$ and $ABDC$ is a parallelogram. Let us describe the translation t on the number plane, using a simple example as an illustration.

Figure 5–27

Let the coordinates of A be $(3, 1)$ and of B be $(-1, 3)$ and let t be the translation such that $t(A) = B$ (see Fig. 5–28). If the coordinates of C are $(5, -4)$, what must be the coordinates of D so that $t(C) = D$, that is, so that D is the image of C under the translation t? Let the coordinates of D be called (d_1, d_2). In Fig. 5–28 two right triangles are formed, $\triangle RAB$ and $\triangle SCD$. Since $\overline{AB} \parallel \overline{CD}$ and $\overline{AR} \parallel \overline{CS}$, we have $\angle RAB \equiv \angle SCD$, and the extra condition that $AB = CD$ tells us that the triangles are congruent. (Why?) Hence $CS = AR$ and $SD = RB$.

172 coordinates and vector geometry

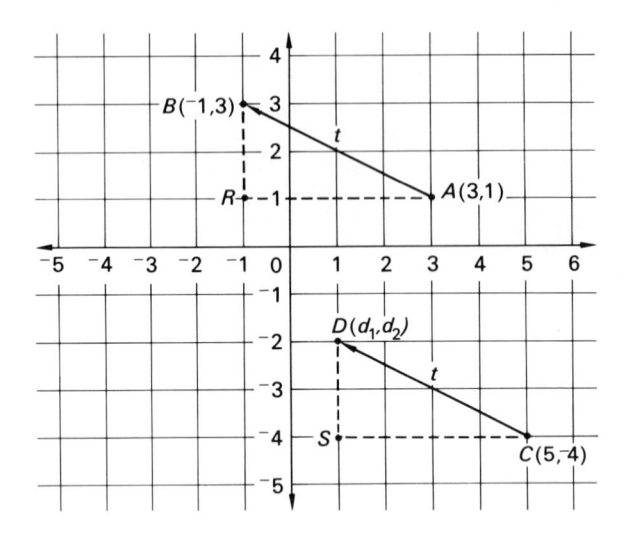

Figure 5–28

From Fig. 5–28 we see that the distance from A to R is $(^-1) - 3$, or $^-4$, and the distance from R to B is $3 - 1$, or 2. Thus the translation t moves the point $(3, 1)$ a distance $|^-4|$ units *to the left* and a distance $|2|$ units *up* to the point $(^-1, 3)$. Since $CS = AR$ and $SD = RB$, the translation t also moves the point $(5, ^-4)$ a distance $|^-4|$ units to the left and a distance $|2|$ units up to the point D. Thus the coordinates of D are $d_1 = 5 + ^-4$ and $d_2 = ^-4 + 2$; that is, D is the point with coordinates $(1, ^-2)$.

What determines this translation t? We know that t moves *each* point $|^-4|$ units to the left and $|2|$ units up. This means that for any point (x, y) in the plane the translation t maps (x, y) onto the point $(x + ^-4, y + 2)$. Thus the translation t is completely determined by the ordered pair of numbers $(^-4, 2)$, and we can call t the translation $[^-4, 2]$, with some special notation such as brackets. Generally:

> If t is the translation $[a, b]$ and if t maps point (u, v) to point (r, s), then $a = r - u$ and $b = s - v$.

After some experience in exploring this numerical description of translations in the plane, it is clear that for each ordered pair of real numbers there is one translation of the plane, and for each translation of the plane there is one ordered pair of real numbers.

Corresponding to each translation of the plane there is a *plane vector:* the set of arrows such that each arrow is directed from a point of the plane to the image of the point under the translation.

> The *plane vector* corresponding to the translation t is the set of all arrows in the plane such that for each point A in $\mathbb{P}$ there is an arrow in the set from point A to point $t(A)$.

Hence the translation t shown in Fig. 5–28, which we call the translation $[^-4, 2]$, is described by the set of all arrows each with its tail at a point (x, y) and its head at point $(x + {}^-4, y + 2)$. In Fig. 5–28 two of these arrows, $\overrightarrow{AB}$, $\overrightarrow{CD}$, are shown. Every arrow in the vector corresponding to $[^-4, 2]$ has its head $|^-4|$ units to the left of its tail and $|2|$ units above its tail.

We see that for our purposes a translation and its corresponding vector are so closely identified that we can safely relax the distinction. We can speak of the vector $[a, b]$, meaning the set of arrows corresponding to the translation $[a, b]$. Thus, for each ordered pair of real numbers (a, b), there is a plane vector $[a, b]$, that is, a translation of the plane: $(x, y) \to (x + a, y + b)$, and for each plane vector (for each translation of the plane) there is one ordered pair of real numbers determined. These correspondences are one-to-one, and we can move freely both in our language and in concept among "translation of the plane," "plane vector," and "ordered pair of real numbers." Thus each arrow in a vector serves to describe the corresponding translation and the corresponding ordered pair of numbers.

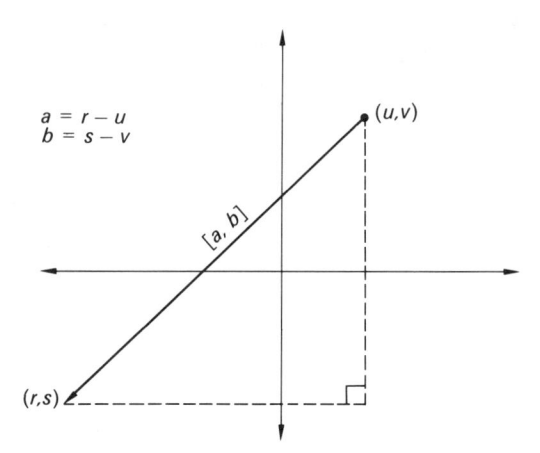

Figure 5–29

Using this relaxed language, we say that the vector $[a, b]$ (see Fig. 5–29) maps each point (x, y) onto the point $(x + a, y + b)$. If a vector t contains an arrow from point (u, v) to point (r, s), then there are real numbers a and b such that

$$r = u + a \quad \text{and} \quad s = v + b,$$

and t is the vector $[a, b]$. (Note that $a = r - u$ and $b = s - v$.)

addition of vectors The composite of two translations is a translation, as we found in Chapter 4. In the language of vectors we call this the *sum* of the vectors, or the vector sum. Our numerical description of a vector allows us to compute the sum of two vectors. For example, consider the vectors $[3, {}^-2]$ and $[{}^-2, 4]$, and let us apply the composite of these vectors (translations) to the point $({}^-4, 3)$:

$$({}^-4, 3) \xrightarrow{[3, {}^-2]} ({}^-4 + 3, 3 + {}^-2) = ({}^-1, 1)$$

and then

$$({}^-1, 1) \xrightarrow{[{}^-2, 4]} ({}^-1 + {}^-2, 1 + 4) = ({}^-3, 5).$$

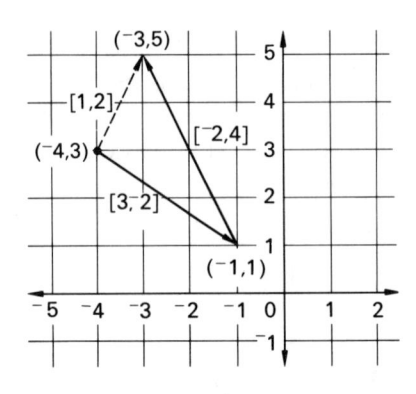

Figure 5–30

Hence the composite translation (vector sum) of $[3, {}^-2]$ and $[{}^-2, 4]$ maps the point $({}^-4, 3)$ onto the point $({}^-3, 5)$ (see Fig. 5–30). That is,

$$({}^-4, 3) \rightarrow ({}^-3, 5),$$

and this mapping is the vector $[1, 2]$, since

$$ {}^-3 - {}^-4 = 1 \quad \text{and} \quad 5 - 3 = 2.$$

We can write this simply as

$$[3, {}^-2] + [{}^-2, 4] = [1, 2].$$

Studying this equality, we can make a simple rule for adding vectors, namely,

$$[a, b] + [c, d] = [a + c, b + d].$$

A proof of this formula follows as we generalize the above example. For any point (x, y) of the plane,

$$(x, y) \xrightarrow{[a, b]} (x + a, y + b)$$

and

$$(x + a, y + b) \xrightarrow{[c, d]} ((x + a) + c, (y + b) + d),$$

so that

$$(x, y) \xrightarrow{[a, b] + [c, d]} (x + (a + c), y + (b + d)).$$

This mapping is the vector $[a + c, b + d]$, and we have the result

$$[a, b] + [c, d] = [a + c, b + d].$$

The addition of vectors allows us to consider each vector as a special sum of vectors. For example, the vector $[3, {}^-2]$ moves each point of the plane $|3|$ units to the right and $|{}^-2|$ units down; the first of these motions is given by the vector $[3, 0]$ and the second by $[0, {}^-2]$, so that

$$[3, {}^-2] = [3, 0] + [0, {}^-2],$$

which agrees with our definition of vector addition. In this way each vector can be expressed as the sum of *component* vectors which move points parallel to the axes of the coordinate system (see Fig. 5–31):

$$[a, b] = [a, 0] + [0, b].$$

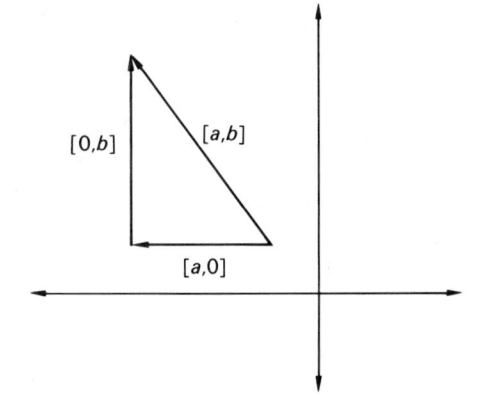

Figure 5–31

The word *add*, when applied to vectors, is not as arbitrary as it may at first seem. We expect certain properties to hold for the addition operation wherever we use it, so we ask whether vector addition has all the properties of a commutative group as defined in Chapter 4. We can show that addition of vectors does indeed have these properties:

> If $[a, b]$, $[c, d]$ and $[e, f]$ are any plane vectors, then
> 1. $([a, b] + [c, d]) + [e, f] = [a, b] + ([c, d] + [e, f])$;
> vector addition is associative.
> 2. $[0, 0] + [a, b] = [a, b] + [0, 0] = [a, b]$;
> there is an identity vector for vector addition, namely, $[0, 0]$.
> 3. $[a, b] + [^-a, ^-b] = [^-a, ^-b] + [a, b] = [0, 0]$;
> for each vector there is an inverse vector such that their sum is $[0, 0]$.
> 4. $[a, b] + [c, d] = [c, d] + [a, b]$;
> vector addition is commutative.

Proofs are left as exercises. (Additional properties of vectors are given in Section 5–4.)

Remarks Since vectors correspond to ordered pairs, the equality of two vectors must imply the equality of their ordered pairs. Thus

> $[a, b] = [c, d]$ if and only if $a = c$ and $b = d$.

For example, if the vectors $[u - 1, 3]$ and $[2, v + 1]$ are equal, we conclude that $u - 1 = 2$ and $3 = v + 1$, that is, $u = 3$ and $v = 2$.

Having developed an operation of addition on vectors, we naturally look for a multiplication operation. There is an operation that can be called vector multiplication, but we are not concerned with it at present. Instead, in the next section we shall "multiply" a number and a vector.

exercise set 5–3

1. The vector t takes the point $(^-1, 2)$ to the point $(4, 3)$. Then
   a) t takes $(^-3, 1)$ to (,)   b) t takes $(4, ^-2)$ to (,)

c) t takes (,) to $(3, {}^-5)$ d) t takes $({}^-\frac{2}{3}, \frac{5}{4})$ to (,)
e) t takes (x, y) to (,)
f) The vector t is determined by the ordered pair of numbers [,].

2. a) The vector $[2, {}^-3]$ maps point $({}^-1, 2)$ onto point (,).
   b) The vector $[{}^-3, 5]$ maps point (,) onto point $(0, {}^-1)$.
   c) The vector [,] maps point $(0, {}^-2)$ onto point $({}^-2, 3)$.

3. a) The vector $[{}^-1, 1]$ maps point $(4, 1)$ onto point (,); then the vector $[1, {}^-2]$ maps point $(3, 2)$ onto point (,). Hence the vector $[{}^-1, 1] + [1, {}^-2]$ maps point $(4, 1)$ onto point (,).
   b) In Fig. 5–32, s is the vector sum of t and r. Then
   $$t = [\ ,\ ],\quad r = [\ ,\ ],\quad s = [\ ,\ ].$$
   c) In Fig. 5–32,
   $(3, 2) \xrightarrow{t} (\ ,\ )$, and $({}^-2, 1) \xrightarrow{r} (\ ,\ )$; then $(3, 2) \xrightarrow{t+r} (\ ,\ )$.

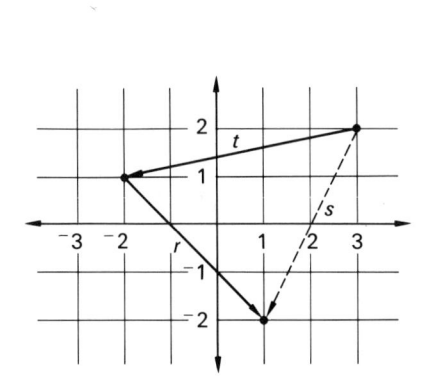

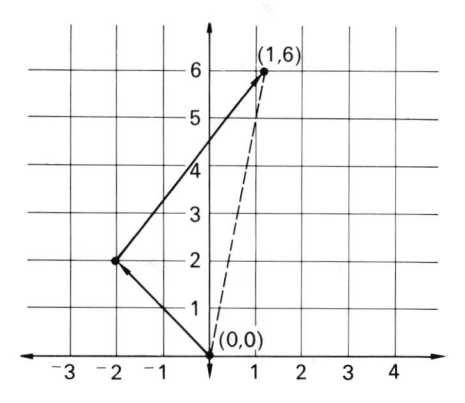

Figure 5–32

Figure 5–33

4. Find the sums of these vectors in two ways: by the formula for vector sum, and by drawing a diagram to show the mapping of point $(0, 0)$ by the composite of the vectors.

Example:

$[{}^-2, 2] + [3, 4] = [{}^-2 + 3, 2 + 4] = [1, 6]$ (see Fig. 5–33).
a) $[{}^-1, {}^-1] + [{}^-3, 4]$ b) $[{}^-7, 2] + [3, {}^-2]$
c) $[{}^-\frac{1}{2}, 2] + [\frac{3}{2}, \frac{1}{2}]$ d) $[3, {}^-\frac{5}{2}] + [{}^-5, \frac{7}{2}]$

5. Find a and b to make these vector equations true.
   a) $[3, a] + [5, {}^-2] = [b, 1]$ b) $[{}^-4, 2] + [a, b] = [3, 5]$
   c) $[{}^-1, 2] + [a, b] = [0, 0]$ d) $[{}^-1, 2] + [a, b] = [{}^-1, 2]$
   e) $[a, {}^-1] + [7, b] = [0, 0]$ f) $[a, 3] + [a, b] = [{}^-2, 2]$

6. Recall that a set with an operation is called a group if (1) the operation is associative, (2) there is an identity element in the set for the operation, (3) every element of the set has an inverse element under the operation. The group is called a commutative group if the operation is commutative.
   a) Do the whole numbers under addition form a commutative group? If not, what properties are lacking?
   b) Do the integers under addition form a commutative group? Explain how the extension from the whole numbers to the integers provides the necessary group properties of addition.
   c) If a and b are any *whole numbers*, can you solve the equation $a + x = b$, for x a whole number? If a and b are any *integers*, can you solve the equation $a + x = b$, for x an integer? Relate your answers to the properties of addition of whole numbers versus addition of integers.

7. a) With arrow diagrams illustrate the four commutative group properties of the set of vectors under addition (see page 176).
   b) Give examples of these four commutative group properties, using specific vectors.
   *c) Prove the four commutative group properties of the vector sum by applying the definition

   $$[a, b] + [c, d] = [a + c, b + d]$$

   and using the fact that the real numbers under addition form a group.

teaching questions and projects 5-1

1. Choose three of the exercises in Exercise Set 5-1 and describe how the same ideas could be presented as puzzles for children.

2. In a spiral curriculum, a mathematical idea is presented early in elementary school and returned to at various later levels, each time extending its scope and sophistication. Describe how the number line could be treated at three or four different levels in a spiral development.

3. Find and organize a set of data that children might use to construct a bar graph; a picture graph.

4. Describe a game which would help children learn to plot points on a number plane.

5. Describe an activity in which children are helped to discover what mathematical sentences have parallel linear graphs.

6. Develop a lesson plan for the development of the properties of addition of vectors, and discuss the student levels at which this could be done.

7. Choose three exercises in Exercise Set 5–3 and explain how the same ideas could be used in exercises for children at a given level.

8. Describe an activity which helps children relate translations as motions in the plane to ordered pairs of real numbers.

9. What are some physical situations illustrating vectors that could be used in problems for children? (For example, forces acting on particular moving objects.)

5–4
the Euclidean plane

A translation of the line moves each point a fixed distance right or left on the line. This distance is the length of the arrow we use to represent the translation. For example, if the 5-point is moved by a translation to the ⁻2-point, the distance is $|{-2} - 5| = |{-7}|$ and we move $|{-7}|$ units to the left.

length of a vector The problem of finding the distance a point is moved in the plane by a vector is solved by using the Pythagorean theorem. For example, how far does the vector [⁻4, 3] move each point of the plane? Let us apply the vector to any point, say (1, 2). Since [⁻4, 3] takes (1, 2) to (⁻3, 5) we can form a right triangle with vertices at (1, 2), (⁻3, 2), and (⁻3, 5), as shown in Fig. 5–34. The lengths of the legs of this right triangle are $|{-4}|$ and $|3|$.

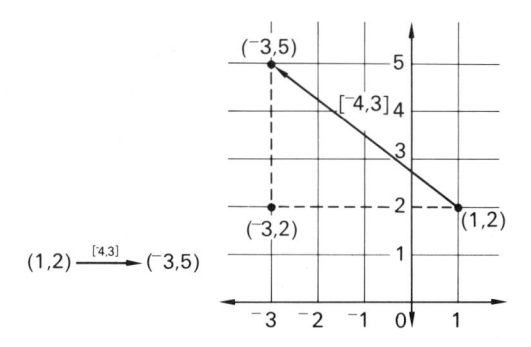

Figure 5–34

Since $|^-4|^2 = (^-4)^2$ and $|3|^2 = 3^2$, the hypotenuse has length $\sqrt{(^-4)^2 + 3^2} = 5$, and we find that the point $(1, 2)$ is moved 5 units to the point $(^-3, 5)$. Since the vector moves each point the same distance, we say that the vector $[^-4, 3]$ has *length* or *magnitude* 5, and we write

$$|[^-4, 3]| = 5.$$

We say, in general, that the *length* of any vector $[a, b]$ is

$$|[a, b]| = \sqrt{a^2 + b^2},$$

which is a real number that is never negative. This means, of course, that each arrow in the vector $[a, b]$ has length $\sqrt{a^2 + b^2}$.

distances between points By definition, a translation preserves lengths of line segments. That is, if translation t maps point A onto B and maps point C onto D, then $AC = BD$. We can use our information about lengths of vectors to give us a formula for distances between points.

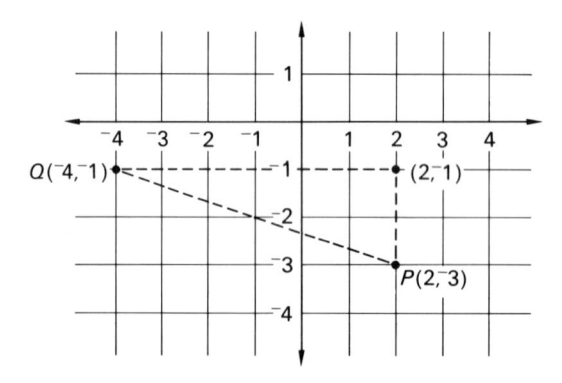

Figure 5–35

For example, consider the points P and Q with coordinates $(2, ^-3)$ and $(^-4, ^-1)$, respectively (Fig. 5–35). We see that the length of $\overline{PQ}$ is the same as the length of the arrow from P to Q. Since $^-4 - 2 = ^-6$ and $^-1 - ^-3 = 2$, this arrow belongs to the vector $[^-6, 2]$, and we know that

$$|[^-6, 2]| = \sqrt{(^-6)^2 + 2^2} = \sqrt{40}.$$

Thus $PQ = \sqrt{40}$. We can extend this procedure to any two points P and Q of the plane with coordinates (p_1, p_2) and (q_1, q_2). Since the vector taking P to Q is $[q_1 - p_1, q_2 - p_2]$, the distance

between P and Q (that is, the length of the arrow from P to Q) is
$$PQ = |[q_1 - p_1, q_2 - p_2]| = \sqrt{(q_1 - p_1)^2 + (q_2 - p_2)^2}.$$

As another example, find the distance between the points $(^-1, 4)$ and $(3, 6)$. Here $(p_1, p_2) = (^-1, 4)$ and $(q_1, q_2) = (3, 6)$, so that the distance between these points is
$$|[3 - {}^-1, 6 - 4]| = \sqrt{4^2 + 2^2} = \sqrt{20}.$$

Using the distance formula, we can verify that translations preserve distances. For example, in Fig. 5–28 the translation $[^-4, 2]$ mapped $A(3, 1)$ onto $B(^-1, 3)$ and mapped $C(5, {}^-4)$ onto $D(1, {}^-2)$. Using the distance formula,
$$AC = |[5 - 3, {}^-4 - 1]| = \sqrt{2^2 + (^-5)^2} = \sqrt{29}$$
and
$$BD = |[1 - {}^-1, {}^-2 - 3]| = \sqrt{2^2 + (^-5)^2} = \sqrt{29}.$$

multiplying a vector by a number Now that we know that every vector has a length, we can stretch or shrink a vector to any given length (provided the vector is not the zero vector $[0, 0]$, which is the only vector with length 0). As an example, let us stretch the vector $[3, 4]$ to twice its length. We know that $|[3, 4]| = \sqrt{25} = 5$, and doubling this gives a vector of length 10. Looking at Fig. 5–36, we suspect that the stretched vector is $[6, 8]$. Checking this, we find that
$$|[6, 8]| = \sqrt{36 + 64} = \sqrt{100} = 10.$$

We doubled the length of the vector by doubling each of the numbers 3 and 4. In the process we also double the length of

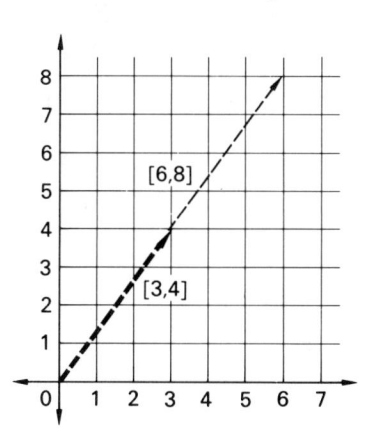

Figure 5–36

each arrow of the vector while leaving the directions of the arrows unchanged. This leads us to try the general case:

Multiply each component of the vector $[a, b]$ by the positive number c, getting the vector $[ca, cb]$. Then:

$$\|[a, b]\| = \sqrt{a^2 + b^2}$$

and

$$\|[ca, cb]\| = \sqrt{c^2 a^2 + c^2 b^2} = \sqrt{c^2} \sqrt{a^2 + b^2} = c\sqrt{a^2 + b^2}.$$

Thus we have shown that $[ca, cb]$ is c times as long as $[a, b]$ if $c > 0$.

Let us consider what happens if c is a negative number, say $c = {}^-2$. In this case, we see that $[{}^-2 \cdot 3, {}^-2 \cdot 4]$ is twice as long as $[3, 4]$, but is "pointed in the opposite direction," or has opposite sense (see Fig. 5–37).

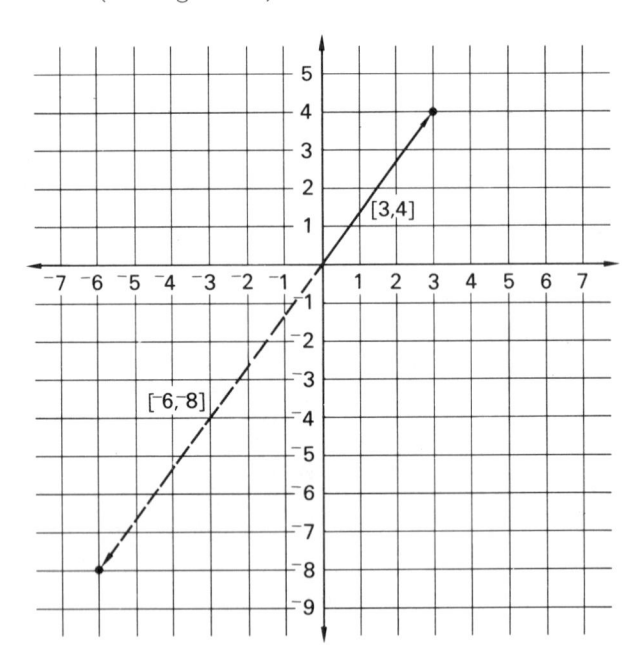

Figure 5–37

We accomplish this stretching or shrinking of a vector in either direction by "multiplying" each component of the vector by a number: a positive number if the vectors have the same sense, a negative number if they have the opposite sense. We can show this with the sentence

$$c[a, b] = [ca, cb].$$

the Euclidean plane With this background we can think of vectors in a somewhat simplified way. For example, consider the vector [⁻3, 5], which moves each point of the plane |⁻3| units to the left and |5| units up. Let us use vector addition (the component form of the vector) to write

$$[^-3, 5] = [^-3, 0] + [0, 5].$$

Then by multiplying vectors by real numbers we show that

$$[^-3, 5] = {^-3}[1, 0] + 5[0, 1].$$

In other words, we can obtain the vector [⁻3, 5] by starting with the vector [1, 0] stretching it |⁻3| units in the opposite direction of [1, 0], then adding the vector [0, 1] after it has been stretched |5| units in the same direction as [0, 1] (see Fig. 5–38).

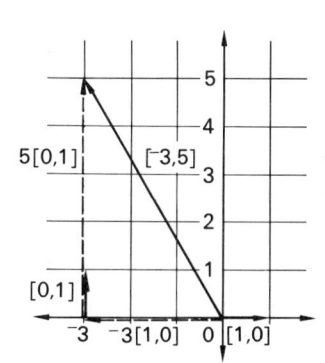

Figure 5–38

For any vector [a, b],

$$[a, b] = a[1, 0] + b[0, 1].$$

This shows that the two vectors [1, 0] and [0, 1] are "basic" in the sense that by stretching, shrinking and adding them we can form any plane vector from them. We say that [1, 0], [0, 1] are a *basis* for the set of plane vectors, and that any vector in the plane is a *linear combination* of [1, 0] and [0, 1].

The process of multiplying a vector by a number (called *scalar multiplication*) has these properties (this is a continuation of the properties of vectors begun in Section 5–3):

If $[a, b]$ and $[c, d]$ are any plane vectors and m and n are any real numbers, then

5. $m([a, b] + [c, d]) = m[a, b] + m[c, d]$.
6. $(m + n)[a, b] = m[a, b] + n[a, b]$.
7. $m(n[a, b]) = (mn)[a, b]$.
8. $1 \cdot [a, b] = [a, b]$.

We complete the list of properties of vectors by including the properties of lengths of vectors:

9. $|[a, b]| \geq 0$, $|[0, 0]| = 0$.
10. $|m| \, |[a, b]| = |[ma, mb]|$.
11. $|[a, b] + [c, d]| \leq |[a, b]| + |[c, d]|$.

By now it is clear that when we know where a vector moves any one point of the plane, we then know where it moves every point. Thus we can simplify all the notation we have used by agreeing to concentrate on the point $(0, 0)$. Since the vector $[a, b]$ moves $(0, 0)$ to (a, b), we can replace the *vector* $[a, b]$ by the *point* (a, b), always remembering that the point represents a vector, namely the vector that maps $(0, 0)$ onto that point.

With this agreement we can think of the eleven properties listed above as describing "addition" of points; "multiplication" of numbers and points; and "lengths" of points [distances from $(0, 0)$ to the points]. This forms an arithmetic of points (ordered pairs of real numbers) of the plane that allows us to do all the mathematics of plane geometry in a systematic numerical way. The plane, with these interpretations of "addition," "multiplication," and "distance" for its points, is called the *Euclidean plane*. Once a person grasps the beginning intuitive feeling for this concept, he is a long way toward a mature view of geometry.

examples in elementary school mathematics programs The approach to these concepts in an elementary school mathematics program (as we continually emphasize) should be at an experimental, intuitive level. Attempts at an early formal development discourage the use of imagination in the development of ideas. Consider this example:

- *Problem* Given a line segment $\overline{AB}$, find its midpoint C.

If left to their own devices, children, depending on their level of maturity, might use any of the following ideas in solving the problem:

1. Trace the line segment on paper. Then fold the paper so that A and B coincide. The point where the folding line intersects $\overline{AB}$ is C. This result follows immediately when we see from the folding that $\overline{AC}$ lies on $\overline{BC}$, and hence $\overline{AC}$ and $\overline{BC}$ are congruent.

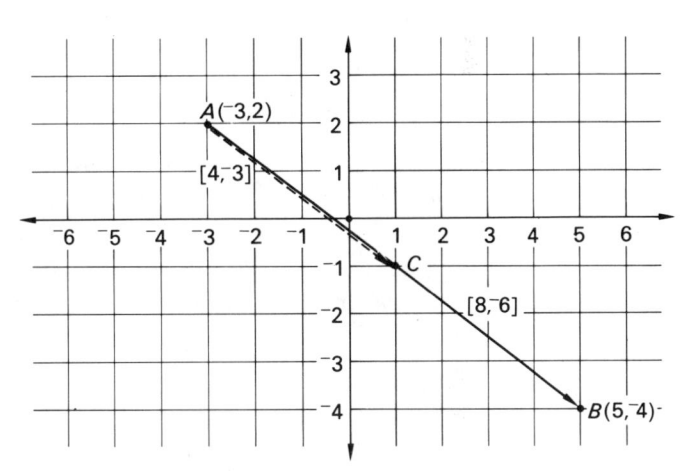

Figure 5-39

2. Let the points A and B be given specific coordinates on the number plane, say $(^-3, 2)$ and $(5, ^-4)$ (Fig. 5–39). Then the point midway between A and B is the point one-half the way from A to B along $\overline{AB}$. The vector taking A to B is the vector $[5 - {^-3}, {^-4} - 2]$, or $[8, {^-6}]$. But we know that the vector taking A to C must be one-half as long, that is, $\frac{1}{2}[8, {^-6}]$ or $[4, {^-3}]$. Then we simply apply the vector $[4, {^-3}]$ to A:

$$(^-3, 2) \xrightarrow{[4, {^-3}]} (1, {^-1}).$$

Thus C is the point $(1, {^-1})$.

3. We can develop a general formula for the coordinates of the midpoint of a segment and apply it to any particular endpoints.

Following the steps in (2), let A be (a_1, a_2), B be (b_1, b_2); then the vector taking A to B is $[b_1 - a_1, b_2 - a_2]$, and one-half this vector is $[\frac{1}{2}(b_1 - a_1), \frac{1}{2}(b_2 - a_2)]$. This vector maps (a_1, a_2) onto $(a_1 + \frac{1}{2}b_1 - \frac{1}{2}a_1, a_2 + \frac{1}{2}b_2 - \frac{1}{2}a_2)$, or $(\frac{1}{2}(a_1 + b_1), \frac{1}{2}(a_2 + b_2))$, which is the midpoint C. Thus the coordinates of the midpoint turn out to be the averages of the coordinates of the endpoints.

Consider another example:

Problem If we join the midpoints of two sides of a triangle, what can we say about the line segment formed, with respect to the third side of the triangle?

If a child were free to use any device and not forced to fit the problem to formal techniques, he might make the decision that the line segment is parallel to and half the length of the third side. He might use any of the following ways of thinking or proceeding, depending on the stage of the school program.

1. Several triangles are drawn and midpoints of two sides of each triangle are found by measuring sides and dividing by two, *or* by folding sides of cutouts of triangles. Then the resulting line segment is compared with the third side of each triangle, first by measuring to find ratios of lengths, and second by measuring to find out whether the lines are parallel. The same information can be found by folding a cutout along median segments (segments joining midpoints of sides), then folding again (as in Fig. 5–40) to see that a rectangle is formed. (How does the rectangle show these results?) Then the conjecture can be made that for all triangles, if the midpoints of two sides are joined, the line segment formed is parallel to and half the length of the third side.

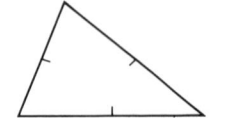

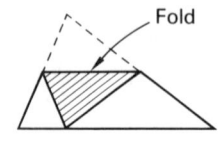

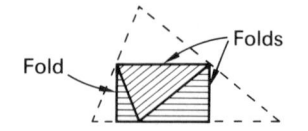

Figure 5–40

2. Draw a triangle on the number plane, with vertices chosen as conveniently as possible, and "see" what the triangle is like. For example, let A, B, C be at $(0, 0)$, $(4, 0)$, $(3, 2)$, respectively. Then the midpoints of $\overline{AB}$ and $\overline{AC}$ can be found by taking $\frac{1}{2}$ of each of the vectors $[4, 0]$ and $[3, 2]$ (that is, we shrink these vectors to one-half their lengths, as in the previous example). The midpoints

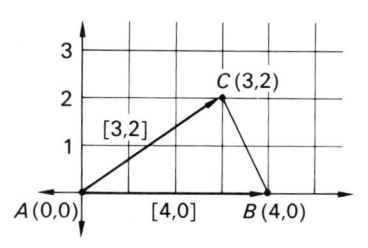

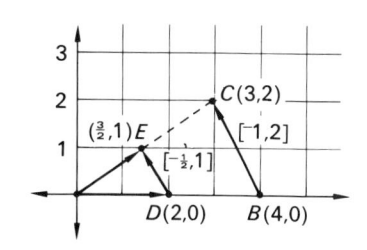

Figure 5-41

D and E are at (2, 0) and ($\frac{3}{2}$, 1), as shown in Fig. 5-41. Finally, compare $\overline{DE}$ with $\overline{BC}$ by comparing the vector taking D to E, [$\frac{3}{2}$ − 2, 1 − 0], with the vector taking B to C, [3 − 4, 2 − 0]. Since [$^-\frac{1}{2}$, 1] = $\frac{1}{2}$[$^-$1, 2], we know that the arrow from D to E is one-half the length of the arrow from B to C, and these arrows are along parallel lines.

3. Choose three points (0, 0), (a, 0), (b, c) as vertices of a triangle on the number plane, for *any* nonzero numbers a, b, c, and repeat (2) to show that the result is true for all triangles.

exercise set 5-4

1. a) A translation t of the number line takes the $^-$3-point to the $^-$10-point. What is the length of the vector t?
   b) A translation r of the number plane takes the ($^-$2, 3)-point to the (4, $^-$5)-point. Then r = [,]. What is the length of the vector r?
   c) Draw a diagram of part (b) and explain how a right triangle is formed, with the vector r on the hypotenuse. What are the lengths of the legs of this triangle?

2. Find:
   a) $\|[^-5, 12]\|$
   b) $\|[^-4, ^-3]\|$
   c) $\|[1, 2]\|$
   d) $\|[9, ^-12]\|$
   e) $\|[^-4, 5]\|$
   f) $\|[\frac{3}{5}, ^-\frac{4}{5}]\|$
   g) Which of these vectors have rational lengths?

3. For x a real number, the symbol $|x|$ has the meaning

$$|x| = x, \text{ if } x \geq 0; \qquad |x| = ^-x, \text{ if } x < 0.$$

Then the function $x \to |x|$ maps each real number onto a non-negative real number.

Consider the symbol

$$\|[a, b]\| = \sqrt{a^2 + b^2}.$$

Here the function $[a, b] \to |[a, b]|$ maps each vector onto its length, a nonnegative real number. Verify with special cases that
a) $|x + y| \leq |x| + |y|$. (Try $x = {}^-3, y = 2; x = 5, y = 1$.)
b) $|[a, b] + [c, d]| \leq |[a, b]| + |[c, d]|$. (Try $[a, b] = [3, {}^-4], [c, d] = [{}^-8, 6]; [a, b] = [4, 3], [c, d] = [{}^-12, {}^-9]$.)
c) Draw a vector diagram illustrating the vectors $[a, b], [c, d]$ and $[a, b] + [c, d]$. Explain the meaning of the inequality in part (b) in terms of the lengths of the sides of a triangle.

*4. In the proof that

$$c[a, b] \text{ is } c \text{ times as long as } [a, b] \text{ if } c > 0,$$

we need to show that

$$|[ca, cb]| = \sqrt{c^2 a^2 + c^2 b^2} = \sqrt{c^2} \sqrt{a^2 + b^2} = c\sqrt{a^2 + b^2}.$$

Prove this result. [*Hint:* Recall that for positive x, t,

$$x^2 = t \quad \text{if and only if} \quad x = \sqrt{t}.$$

Use this to show that $\sqrt{rs} = \sqrt{r}\sqrt{s}$, if $r > 0, s > 0$. Then let $r = c^2, s = a^2 + b^2$.]

5. a) $|[{}^-3, 4]| = x; \ 3|[{}^-3, 4]| = y; \ |[{}^-9, 12]| = z$.
   Find x, y, and z and describe how to stretch a vector to 3 times its length.
   b) $|[{}^-4, 3]| = x; \ \frac{1}{5}|[{}^-4, 3]| = y; \ |[{}^-\frac{4}{5}, \frac{3}{5}]| = z$.
   Find x, y, and z and describe how to shrink a vector to $\frac{1}{5}$ its length.

6. Let s be a function that maps each point (x, y) of the plane to the point $(3x, 3y)$. Then

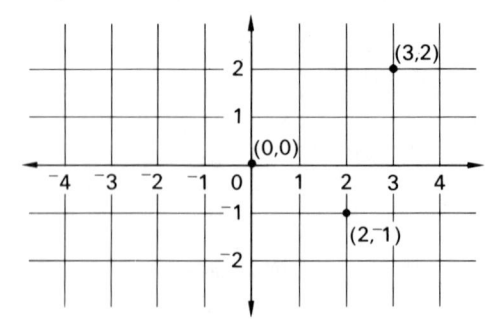

Figure 5–42

a) $(0, 0) \xrightarrow{s} (\ , \ ); \ (3, 2) \xrightarrow{s} (\ , \ ); \ (2, {}^-1) \xrightarrow{s} (\ , \ )$
b) The triangle A with vertices at $(0, 0), (3, 2), (2, {}^-1)$ is taken by s to the triangle B. What are the vertices of triangle B?

c) Are these two triangles congruent?
d) Do these two triangles have the same shape?
e) Find the lengths of the sides of triangles A and B. Find the three ratios, each of which is the ratio of the length of a side of A to the length of the corresponding side of B. What is true of these three ratios? [*Hint:* $\sqrt{117} = \sqrt{13} \cdot \sqrt{9}$; see Exercise 4.]

7. Determine the vectors:
   a) $3[1, {}^-2] + 4[0, 3]$
   b) $2[3, {}^-1] - \frac{1}{2}[2, {}^-5]$
   c) $\frac{1}{2}[2, {}^-3] - \frac{2}{3}[3, {}^-2]$

8. Show how to stretch and shrink the vectors $[1, 0]$ and $[0, 1]$ and to add the results, in order to get the vectors:
   a) $[{}^-3, {}^-4]$
   b) $[5, {}^-3]$
   c) $[\frac{4}{3}, {}^-\frac{2}{5}]$

9. Find a, b (if possible), so that each of these vector equations is true.
   Example: $[{}^-3, 5] = a[1, 1] + b[1, {}^-1]$.
   $$a[1, 1] + b[1, {}^-1] = [a, a] + [b, {}^-b] = [a + b, a - b].$$
   Then $[{}^-3, 5] = [a + b, a - b]$ if $a + b = {}^-3$ and $a - b = 5$; that is, if $a = 1$ and $b = {}^-4$.
   a) $2[a, {}^-2] + 3[1, b] = [0, 0]$
   b) $a[{}^-3, 1] + [2, {}^-2] = [{}^-2, b]$
   c) $a[1, 2] + b[0, {}^-2] = [2, 4]$
   d) $[3, {}^-1] = a[1, 0] + b[1, 1]$
   e) $a[2, {}^-1] + b[{}^-4, 2] = [0, 2]$
   f) $a[1, 1] + b[{}^-1, 1] = [2, 1]$
   g) $a[1, {}^-1] + b[{}^-1, 1] = [2, {}^-1]$

10. Two vectors are called *parallel* if one is a scalar multiple of the other. That is, $[c, d]$ is parallel to $[e, f]$ if and only if there is a real number t such that $[c, d] = t[e, f]$ or $t[c, d] = [e, f]$. (Reasons for this definition will be discussed in the next section.)
    a) Which of the following pairs of vectors are parallel?
       i) $[2, {}^-3], [1, {}^-3]$
       ii) $[2, {}^-3], [{}^-6, 9]$
       iii) $[{}^-1, 1], [1, {}^-1]$
       iv) $[0, 0], [5, 1]$
    b) Can a pair of parallel vectors form a basis for the set of plane vectors? [*Hint:* Consider Exercise 9(e) and (g).]
    c) If two vectors are not parallel, do they form a basis for the set of plane vectors? (Argue in terms of stretching, shrinking, and adding two nonparallel vectors to obtain any given vector.)

11. Verify the properties of scalar multiplication and of lengths of vectors for the specific vectors $[a, b] = [{}^-3, 4], [c, d] = [2, {}^-3]$ and the numbers $m = {}^-3, n = 2$. (See properties 5 through 11 on page 184.)

coordinates and vector geometry

*12. Prove scalar multiplication properties 5 through 8, using the definitions
$$[a, b] + [c, d] = [a + b, c + d],$$
$$m[a, b] = [ma, mb].$$

In your proofs, you must distinguish between two different uses of the symbol "+," one for vectors and the other for numbers. Point out where this distinction is needed.

13. a) Explain how Fig. 5–43 enables us to find the coordinates of the midpoint M of the line segment joining the points (2, 3) and (6, 1). (Note that the vector from (2, 3) to (6, 1) is [4, ⁻2].)

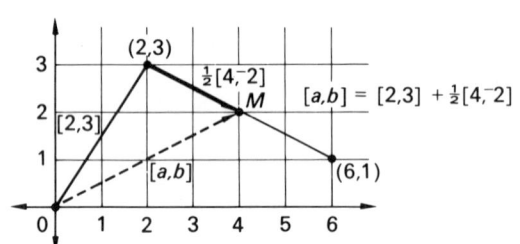

Figure 5–43

b) Use the same procedure to find the coordinates of the midpoint of the segment joining

i) (⁻2, 3) and (4, ⁻5), ii) (3, 2) and (⁻4, ⁻5).

c) Show that, in general, each coordinate of the midpoint is the average of the corresponding coordinates of the endpoints.

14. a) In the triangle (Fig. 5–44) with vertices at X, (0, 0); Y, (6, 0); Z, (2, 4), find the midpoints A and B and show that the segment $\overline{AB}$ is parallel to and half the length of segment $\overline{XY}$ (see Exercises 10 and 13 above).

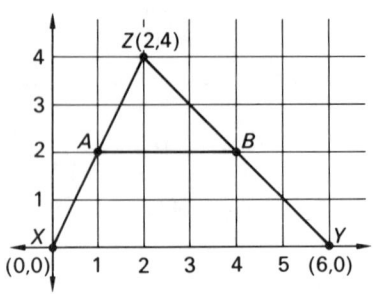

Figure 5–44

b) Repeat part (a) for the triangle X, (0, 0); Y, (3, ⁻2); Z, (⁻1, 4). Does the result still hold true?

c) Prove this result for any triangle.

5-4   the Euclidean plane   191

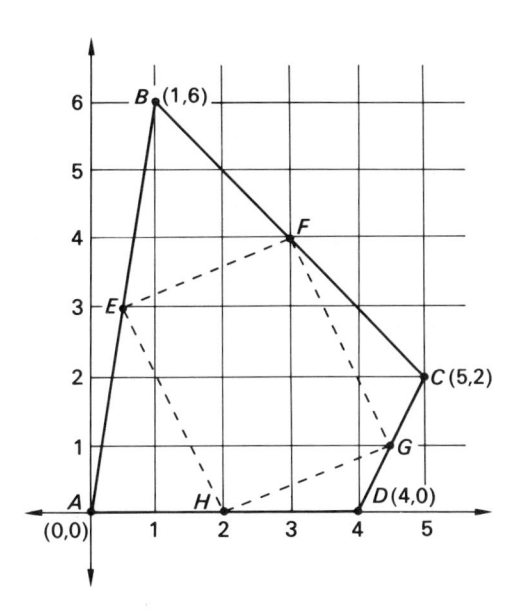

Figure 5–45

15. a) Show that the midpoints of the sides of the quadrilateral (Fig. 5–45) with vertices at (0, 0), (4, 0), (5, 2), (1, 6) are the vertices of a parallelogram.
    b) Show that this is true for any quadrilateral.

16. Look at Fig. 5–46 and explain how it helps us to find the coordinates of the point T that is $\frac{1}{3}$ of the way from (2, 4) to (¯3, 1). (Note that the vector taking (2, 4) to (¯3, 1) is [¯5, ¯3], and hence the vector taking (2, 4) to T is $\frac{1}{3}$[¯5, ¯3].)
    a) Use a similar procedure to find the coordinates of:
        i) The point $\frac{2}{3}$ the way from (¯1, 2) to (5, 5).
        ii) The point $\frac{3}{4}$ the way from (1, ¯3) to (¯3, 4).

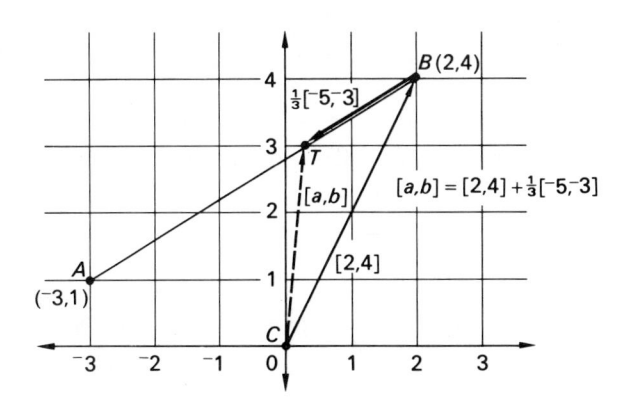

Figure 5–46

*b) Show that, in general, the point that is $\frac{p}{q}$ of the way from (a, b) to (c, d) has coordinates $\left(a + \frac{p}{q}(c - a), b + \frac{p}{q}(d - b)\right)$.

17. Find the midpoints of the sides of the triangle (Fig. 5–47) with vertices at $(0, 0)$, $(8, 0)$, $(2, 6)$ and draw the segment from each vertex to the midpoint of the opposite side. Show that these three segments meet in one point T that is $\frac{2}{3}$ the way from each vertex to the opposite midpoint. (Use the result of Exercise 16.)

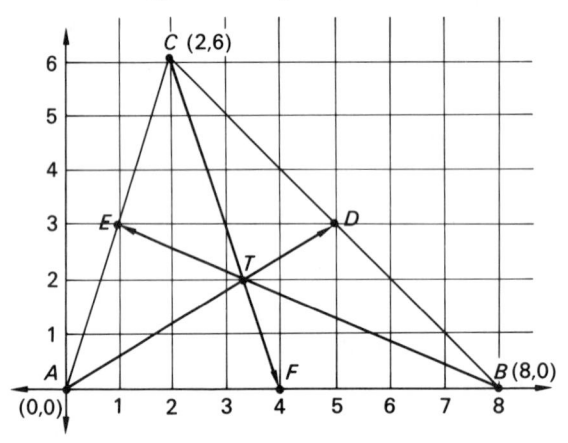

Figure 5–47

5–5
lines in the Euclidean plane

Earlier we showed that the vector $t[a, b]$ is $|t|$ times the length of the vector $[a, b]$ and that it has the same direction as $[a, b]$ if $t > 0$, or it has the opposite direction if $t < 0$. The multiplier t "stretches" or "shrinks" the vector $[a, b]$, but does not change the direction of the lines along which $[a, b]$ lies. For example, the set of arrows shown in Fig. 5–48 all represent the vector $[4, 2]$; whereas the set of arrows shown in Fig. 5–49 all represent the vector $(-\frac{1}{2})[4, 2]$. Since all arrows that represent a nonzero vector $[a, b]$ are parallel to all arrows that represent $t[a, b]$, $t \neq 0$, this suggests a definition of *parallel vectors:*

Two vectors [a, b] and [c, d] are *parallel* if and only if there is a real number t such that [a, b] = t[c, d]; that is, if one vector is a multiple of the other.

5-5 lines in the Euclidean plane 193

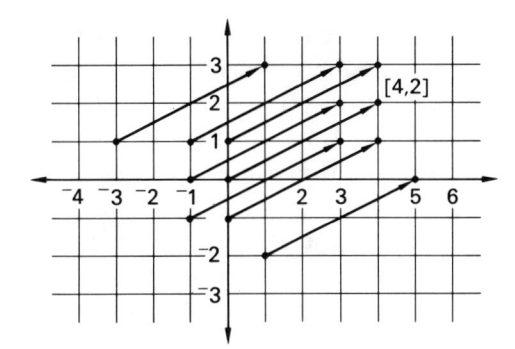

Figure 5–48

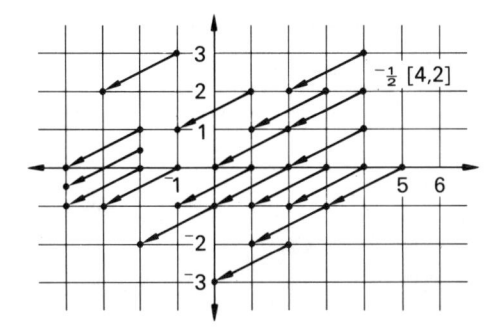

Figure 5–49

For example, $[\frac{2}{3}, {}^-3]$ and $[{}^-2, 9]$ are parallel since

$${}^-3[\tfrac{2}{3}, {}^-3] = [{}^-3 \cdot \tfrac{2}{3}, {}^-3 \cdot {}^-3] = [2, 9].$$

Note that the zero vector $[0, 0]$, by our definition, is parallel to every vector, since $0[c, d] = [0, 0]$ for any c, d.

Furthermore, all the arrows belonging to a given nonzero vector lie on parallel lines, and we say they all have the same *direction*. The direction of a line is then characterized by any nonzero vector whose arrow lies on that line. For example, the

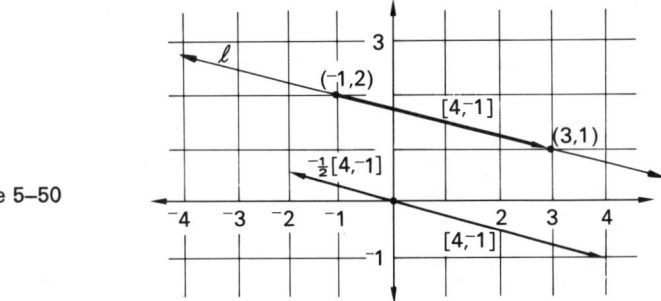

Figure 5–50

line through the points (⁻1, 2) and (3, 1) has direction given by the vector [3 − ⁻1, 1 − 2], since this *direction vector*, [4, ⁻1], has an arrow from (⁻1, 2) to (3, 1) (see Fig. 5–50). It also has direction given by any multiple of [4, ⁻1], such as $-\frac{1}{2}$[4, ⁻1], 3[4, ⁻1].

In order to describe a given line in the plane, we need to specify not only its direction but also some point on the line. Once we have described a line by a point and a direction, we can use algebra of vectors to represent the line by means of an equation. For example, consider the line *l* through point (⁻2, 3) with direction given by the vector [3, ⁻1]. Let (*x*, *y*) be the coordinates of any point of *l*. Our problem is to find an equation in *x* and *y* satisfied by points of *l* and only points of *l*. The vector taking (0, 0) to (⁻2, 3) is [⁻2, 3], called the *position vector* of point (⁻2, 3).

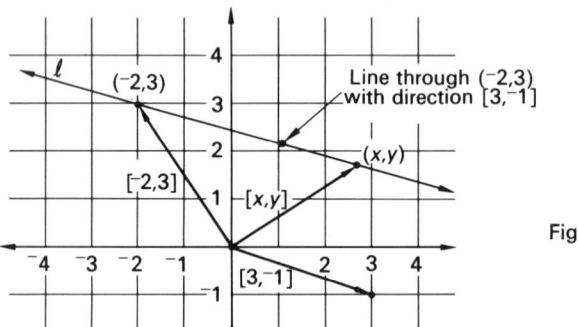

Figure 5–51

The position vector of any point (*x*, *y*) on *l* is [*x*, *y*]. The key observation to make here is that the vector taking (⁻2, 3) on *l* to any other point (*x*, *y*) on *l* is *parallel* to [3, ⁻1]. We calculate the vector taking (⁻2, 3) to (*x*, *y*) as [*x* + 2, *y* − 3]; since this vector is parallel to [3, ⁻1], we conclude that for each point (*x*, *y*) on *l* there is a real number *t* and for each real number *t* there is a point (*x*, *y*) on *l* such that

$$[x + 2, y - 3] = t[3, {}^-1] \qquad \text{(vector equation of } l\text{)}.$$

This vector equation of *l* can quickly be converted to a more familiar equation of *l* by recalling that *t*[3, ⁻1] = [3*t*, ⁻*t*], and that vectors are equal if their corresponding components are equal. Then

$$[x + 2, y - 3] = [3t, {}^-t] \Leftrightarrow x + 2 = 3t \text{ and } y - 3 = {}^-t$$
$$\Leftrightarrow x = {}^-2 + 3t \text{ and } y = 3 - t.$$

These equations of l specify all the points of l. For each real number t there is a point (x, y) on l such that

$$x = {}^-2 + 3t \quad \text{and} \quad y = 3 - t \qquad \text{(parametric equations of } l\text{)}.$$

In particular, for $t = 0$, we have the point $({}^-2, 3)$, the original point specified on l. For $t = 1$, we have the point $(1, 2)$; for $t = {}^-2$, the point $({}^-8, 5)$; etc. The number t in the equations is called the *parameter* in the equations. Note the roles of the given point $({}^-2, 3)$ and the given direction vector $[3, {}^-1]$ in the parametric equations.

We need only eliminate t from the parametric equations of l to obtain the usual form: If $y = 3 - t$, then $t = 3 - y$. Thus $x = {}^-2 + 3t$ becomes $x = {}^-2 + 3(3 - y)$ or $x = 7 - 3y$. We can write this as

$$x + 3y = 7 \qquad \text{(general form of equation of } l\text{)},$$

or

$$y = {}^-\tfrac{1}{3}x + \tfrac{7}{3} \qquad \text{(explicit form of equation of } l\text{)}.$$

The latter of these forms displays the function f given by $f(x) = {}^-\tfrac{1}{3}x + \tfrac{7}{3}$, so that $y = f(x)$ is an equation of l. It also displays what is called the *slope* of l as the coefficient, $^-\tfrac{1}{3}$, of x in the explicit form. Note that the slope of l gives the ratio of the second to the first components of the direction $[3, {}^-1]$ of l.

Of all the various forms of equations of l, the vector and parametric forms preserve the most information about l, namely that l contains the point $(2, {}^-3)$ and is parallel to the vector $[3, {}^-1]$. An example will illustrate this. A line k has the general equation

$$3x - 2y = 8$$

and parametric equations

$$x = 2 - 2t \quad \text{and} \quad y = {}^-1 - 3t.$$

The form of the general equation does not exhibit any specific point that k contains or the direction of k. But the form of the parametric equations for k immediately shows that $(2, {}^-1)$ is a point on k, and the direction of k is given by $[{}^-2, {}^-3]$. To verify this we first set $t = 0$ and find that $(2, {}^-1)$ is a point on k. Then we write the parametric equations for k in the form

$$x - 2 = {}^-2t \quad \text{and} \quad y - {}^-1 = {}^-3t,$$

which means that in vector form
$$[x - 2, y - {}^-1] = [{}^-2t, {}^-3t] = t[{}^-2, {}^-3],$$
since vectors are equal if and only if their respective components are equal. But this vector equation specifies that the vectors from point $(2, {}^-1)$ on k to any other point (x, y) on k be parallel to the vector $[{}^-2, {}^-3]$. Thus k is parallel to $[{}^-2, {}^-3]$.

The generalization of the preceding discussion leads to the various forms of any line in the plane:

> If line k contains point (p, q) and has direction $[a, b]$ (that is, has direction given by the nonzero vector $[a, b]$), then there is a one-to-one correspondence between the points (x, y) of k and the real numbers t such that
> $$[x, y] = [p, q] + t[a, b] \qquad \text{(vector equation of } k\text{)};$$
> that is,
> $$x = p + ta \text{ and } y = q + tb \qquad \text{(parametric equations of } k\text{)}.$$

In the study of lines in the plane, we often describe the direction of a line by the number $\frac{b}{a}$, $a \neq 0$, which we called the slope of the line, where $[a, b]$ is a vector giving the direction of the line. For example, if line k contains the points $({}^-2, 1)$ and $(3, {}^-3)$, then a direction vector of k is $[3 - {}^-2, {}^-3 - 1]$, or $[5, {}^-4]$ (see Fig. 5–52). Thus the slope of k is $-\frac{4}{5}$, which can be interpreted as a number describing the ratio of "the rise to the run" from one point on k to another.

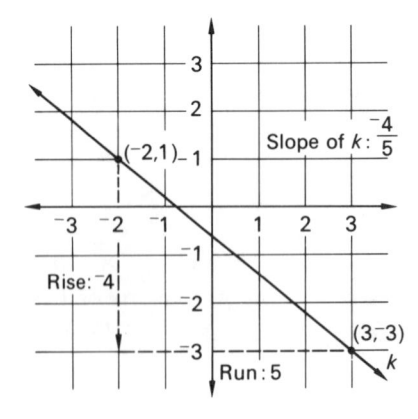

Figure 5–52

We see that "the" slope of a line does not depend on the particular pair of points chosen on the line. If line l contains the

5-5   lines in the Euclidean plane

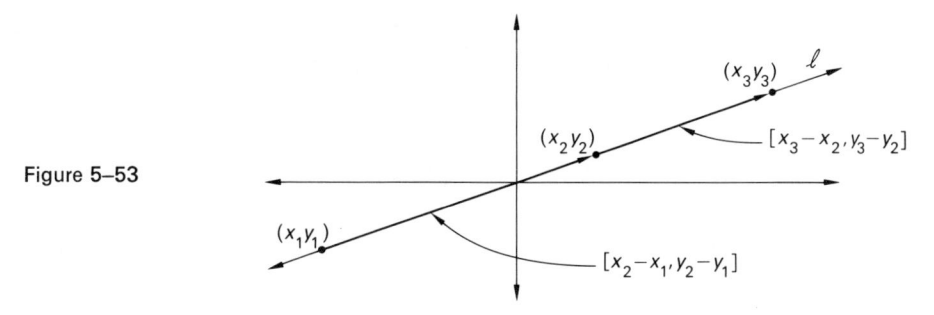

Figure 5-53

distinct points (x_1, y_1), (x_2, y_2), (x_3, y_3), then $[x_2 - x_1, y_2 - y_1]$ and $[x_3 - x_2, y_3 - y_2]$ are two vectors giving the direction of l. Since these two vectors are parallel and nonzero, there is a nonzero number t such that

$$[x_2 - x_1, y_2 - y_1] = t[x_3 - x_2, y_3 - y_2],$$

or, when we equate corresponding coefficients,

$$x_2 - x_1 = t(x_3 - x_2) \quad \text{and} \quad y_2 - y_1 = t(y_3 - y_2).$$

Thus if $x_2 \neq x_1$ and $x_3 \neq x_2$, we obtain

$$\frac{y_2 - y_1}{x_2 - x_1} = \frac{y_3 - y_2}{x_3 - x_2}.$$

Now we can safely define *the* slope of line l.

The slope of line l containing points (x_1, y_1) and (x_2, y_2) is the number

$$\frac{y_2 - y_1}{x_2 - x_1} \quad \text{if } x_1 \neq x_2.$$

A line for which $x_1 = x_2$ is called *vertical*, and we say that it has no slope or that a slope is not defined for this line. The vector equation of the vertical line through the point (p, q) is $[x, y] = [p, q] + t[0, b]$.

Example 1. Find the general and explicit forms of equations for the line through the points $(^-2, 1)$ and $(3, ^-3)$.

The slope of this line is

$$\frac{^-3 - 1}{3 - ^-2} = \frac{^-4}{5}$$

(since a direction vector of the line is the vector $[5, {-4}]$ taking $({-2}, 1)$ to $(3, {-3})$). Since (x, y) and $({-2}, 1)$ are also a pair of points on the line, its slope is also given by

$$\frac{y - 1}{x - {-2}}.$$

Thus

$$\frac{y - 1}{x + 2} = \frac{{-4}}{5},$$

from which we obtain

$$y - 1 = {-\tfrac{4}{5}}(x + 2),$$
$$y = {-\tfrac{4}{5}}x - \tfrac{3}{5} \qquad \text{(explicit form)},$$
$$4x + 5y = {-3} \qquad \text{(general form)}.$$

Example 2. Find the slope of the line whose general equation is $3x + 2y = 6$.

Note that we may write this equation in the forms

$$3x = {-2}(y - 3), \qquad \frac{y - 3}{x} = \frac{{-3}}{2}, \quad \text{and} \quad \frac{y - 3}{x - 0} = \frac{{-3}}{2},$$

which tell us that the points (x, y) and $(0, 3)$ are on the line and that the slope of the line is $-\tfrac{3}{2}$.

Generalizing this example, the general equation $ax + by = c$, with $b \neq 0$, can be written

$$by - c = -ax, \qquad b\!\left(y - \frac{c}{b}\right) = -ax, \quad \text{and} \quad \frac{y - \frac{c}{b}}{x - 0} = \frac{{-a}}{b},$$

which shows that the slope of the line is $\frac{-a}{b}$ and the line contains the point $\left(0, \frac{c}{b}\right)$. On the other hand, writing the equation in explicit form,

$$by = -ax + c, \qquad y = \frac{-a}{b}x + \frac{c}{b},$$

we compare forms and see that the slope is given by the coefficient of x and that the constant identifies the point $\left(0, \frac{c}{b}\right)$.

exercise set 5-5

1. For each pair of points, find a direction vector of the line through the points.
   a) $(^-1, 3)$ and $(4, ^-4)$
   b) $(\frac{1}{3}, \frac{3}{5})$ and $(^-\frac{2}{3}, \frac{4}{5})$
   c) $(\frac{1}{2}, ^-\frac{1}{2})$ and $(1, ^-\frac{3}{2})$
   d) $(^-\frac{1}{3}, ^-\frac{1}{4})$, $(^-\frac{1}{3}, \frac{1}{2})$

2. For each of the following lines, find (i) the vector equation of the line, (ii) the parametric equations of the line, and (iii) a general form of the equation.
   a) Line l through points $(^-1, 3)$ and $(4, ^-4)$.
   b) Line m through point $(\frac{1}{2}, ^-\frac{1}{2})$ with direction vector $[1, ^-2]$.
   c) Line n through points $(\frac{1}{3}, \frac{3}{5})$ and $(^-\frac{2}{3}, \frac{4}{5})$.
   d) Line p through point $(^-\frac{1}{3}, \frac{1}{2})$ with direction vector $[0, 3]$.

3. a) Is point $(3, ^-4)$ on the line with parametric equations $x = 2 - 3t$ and $y = ^-1 + 9t$? If so, what value of t corresponds to this point?
   b) Is point $(3, ^-4)$ on the line with general equation $3x - 4y = 24$?

4. Find, for each of the following lines, a point on the line, a direction vector of the line, and the slope of the line (if it has a slope).
   a) The line q with parametric equations $x = \frac{1}{3} - 2t$, $y = ^-\frac{2}{3} + 3t$.
   b) The line r with vector equation $[x, y] = [^-2, 1] + t[0, ^-3]$.
   c) The line v with general equation $3x + 4y = 12$.
   d) The line u with explicit equation $y = ^-\frac{2}{3}x + 3$.

5. Find (i) a general equation and (ii) parametric equations of a line
   a) through $(^-1, 1)$ with slope 5,
   b) through $(^-2, 1)$ with slope $\frac{1}{5}$.

6. Draw the graphs of the lines described in Exercises 2, 4, and 5.

teaching questions and projects set 5-2

1. Develop a set of puzzles relating to stretching and shrinking vectors.

2. What are some physical models that might be used to illustrate stretching and shrinking vectors?

3. How can children be helped to relate the properties of scalar multiplication to the familiar properties of arithmetic?

4. Choose three exercises from Exercise Set 5–5 and explain how the same ideas could be used in exercises for children at a given level.
5. Describe an activity that could be used to develop the idea of parallel vectors.
6. Find examples in science materials for children to which the concept of vectors would apply.

chapter six

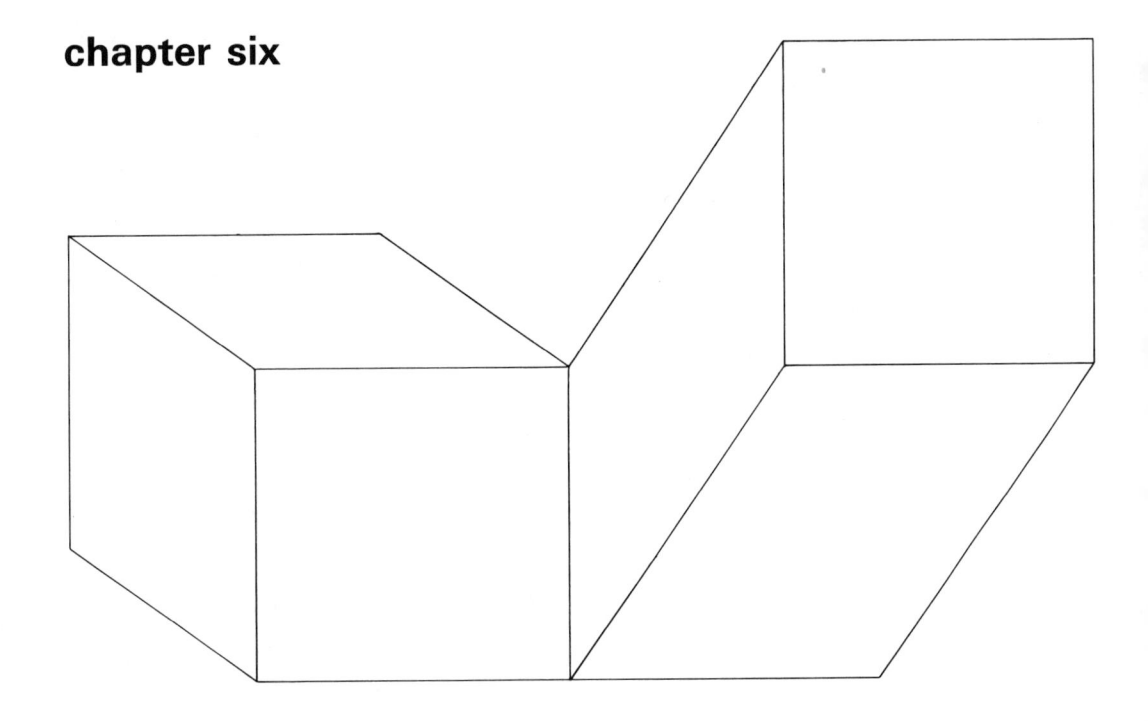

geometric transformations

A dynamic view of geometry, in which we study the properties of figures that remain unchanged when we "move" or "stretch" the plane, is developed in stages. At first, we exploit our intuition by imagining that figures move about in the plane; we move tracings of figures and note relations between their initial and terminal positions. In this way we develop our intuition about rigid motions and symmetries of figures (Chapter 1). Later we reconcile our intuitive notion of moving figures with the fact that figures are fixed subsets of the points of the plane. We do this by describing a plane motion as a function from the set of all points $\mathbb{P}$ constituting the plane to this same set $\mathbb{P}$ (Chapter 4). Functions of the plane are sometimes called *mappings* of the plane or *transformations* of the plane.

6–1
functions of the plane in terms of coordinates

In Chapter 5 we introduced a coordinate system in the plane. Using the resulting number plane, we have an efficient notation for examining functions of the plane more carefully. For example, we established that every translation of the plane is a function from $\mathbb{P}$ to $\mathbb{P}$ which maps each point (x, y) of $\mathbb{P}$ onto the point $(x + a, y + b)$, for some real numbers a and b. A translation is denoted by $[a, b]$ (see Fig. 6–1):

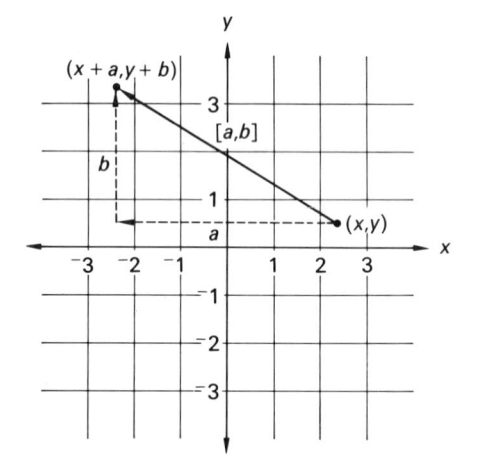

$$(x, y) \xrightarrow{[a, b]} (x + a, y + b).$$

Figure 6–1

Using this notation it is easy to verify that a translation is 1–1 and onto.

Onto: We show that for each point (c, d) of the plane there is a point that is mapped by the translation $[a, b]$ onto (c, d). This is the point $(c - a, b - d)$, since

$$(c - a, d - b) \xrightarrow{[a, b]} (c - a + a, d - b + b) = (c, d).$$

1–1: If there are two distinct points of the plane, say (c, d) and (e, f), that are mapped by $[a, b]$ onto the same point, then

$$(c, d) \xrightarrow{[a, b]} (c + a, d + b), \qquad (e, f) \xrightarrow{[a, b]} (e + a, f + b),$$

and $(c + a, d + b) = (e + a, f + b)$, which would imply that

$$c + a = e + a \quad \text{and} \quad d + b = f + b.$$

But then $c = e$ and $d = f$, violating the assumption that (c, d) and (e, f) are distinct points.

A translation is a 1–1, onto function from $\mathbb{P}$ to $\mathbb{P}$, and each translation t has an inverse t^{-1} such that the composites $t \circ t^{-1}$ and $t^{-1} \circ t$ are the identity function (which leaves every point of $\mathbb{P}$ fixed). For example, the inverse of the translation $[a, b]$ is the translation $[^-a, ^-b]$,* since the composite of $[a, b]$ and $[^-a, ^-b]$ is the translation $[0, 0]$. (Recall that the composite of two translations corresponds to the vector sum of the corresponding vectors.)

We can prove that any 1–1, onto function has an inverse.

▶ *Theorem 1*: If f is a 1–1, onto function from a set S to S, then there is a function g from S to S such that

$$f \circ g = g \circ f = I \quad \text{(the identity function on } S\text{)};$$

furthermore, g is 1–1 and onto.

Proof: Given any element b in S, since f is onto there is an element a in S such that $f(a) = b$. Since f is 1–1, this element a is unique. Define a new function g from S to S as follows:

$$a = g(b) \quad \text{if and only if} \quad b = f(a).$$

Then, for any a in S,

$$g \circ f(a) = g(f(a)) = g(b) = a,$$

and for any b in S,

$$f \circ g(b) = f(g(b)) = f(a) = b,$$

showing that

$$g \circ f = I \quad \text{and} \quad f \circ g = I.$$

The reader is invited in the exercises to prove that if f and g are functions from S to S such that $f \circ g = I$ and $g \circ f = I$, then f and g are 1–1 and onto.

A translation then is a 1–1, onto function of the plane with an inverse. There are functions that are not 1–1 and onto and that do not have inverses. For example, consider the function p which maps each point (x, y) of $\mathbb{P}$ onto the point $(x, 0)$. This function is neither 1–1 nor onto. It maps all the points of $\mathbb{P}$ onto the x-axis. In particular, $(2, b) \xrightarrow{p} (2, 0)$ for all numbers b, so that, for example, every point on the vertical line $x = 2$ is mapped onto the single point $(2, 0)$, as shown in Fig. 6–2.

---

* The symbol "^-x" denotes the additive inverse of x.

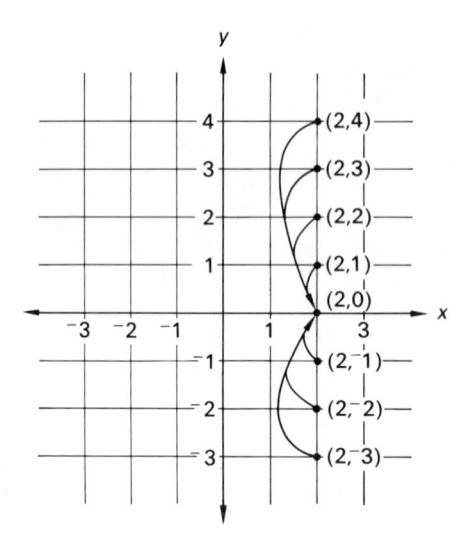

Figure 6–2

6–2
transformations of the plane

The functions of the plane that are most interesting in this study of geometry are those which have inverses. They are called *invertible* or *nonsingular transformations*. We shall call them simply transformations.

▶ *Definition:* An (invertible) *transformation* of the plane is a 1–1, onto function from $\mathbb{P}$ to $\mathbb{P}$.

The fact that a transformation has an inverse allows us to characterize the set T of all transformations of $\mathbb{P}$, under the operation of composition of functions, as a group:

> If f and g are transformations of $\mathbb{P}$, then $f \circ g$ is a transformation of $\mathbb{P}$.
>
> If f, g, and h are transformations, then $(f \circ g) \circ h = f \circ (g \circ h)$.
>
> There is an identity transformation I of $\mathbb{P}$ such that for any transformation f of $\mathbb{P}$, we have $f \circ I = f$ and $I \circ f = f$.
>
> For each transformation f of $\mathbb{P}$ there is a transformation f^{-1} of $\mathbb{P}$ such that $f \circ f^{-1} = f^{-1} \circ f = I$.

Among this group of transformations of $\mathbb{P}$ there is a subset that we considered in Chapter 4: the set E of functions of $\mathbb{P}$ that preserve lengths of line segments. We called these functions *rigid motions*, and we identified some of them as translations, rotations, and reflections. Each of these functions was found to have an inverse, and we accepted without proof the fact that every rigid motion is a translation, a rotation, a reflection or a composite of some of these. Thus we may say that the set E of rigid motions of $\mathbb{P}$ is a subset of the group of transformations of $\mathbb{P}$. We may say even more: the subset E of rigid motions is a *subgroup* of the group of transformations, since this subset has all the properties of a group. The rigid motions are also called *isometries* (*iso*, meaning equal; *metre*, meaning length).

There are transformations that are not isometries. Thus:

The set E of isometries form a *proper* subgroup of the group of transformations.

For example, the function q that maps each point (x, y) of $\mathbb{P}$ onto $(2x, y)$ is invertible (what is q^{-1}?), but is not an isometry. [For example, q maps the segment from $(0, 0)$ to $(1, 0)$, of length 1, onto the segment from $(0, 0)$ to $(2, 0)$, of length 2.]

Isometries preserve lengths of segments. Another important type of transformation of $\mathbb{P}$ preserves parallelism of lines; such a transformation is called a *dilation* of $\mathbb{P}$.

▶ *Definition:* A transformation d of $\mathbb{P}$ is a dilation if for any two distinct points P and Q, segment $\overline{PQ}$ is parallel to segment $\overline{d(P)\,d(Q)}$.

It is clear that if f and g are dilations of $\mathbb{P}$, then $f \circ g$ is also a dilation: let g map line l onto k so that $k \parallel l$, and let f map line k onto line m such that $m \parallel k$; then $f \circ g$ maps l onto m, and $m \parallel l$. We see that dilations, being functions, are associative under composition; also the identity function is a dilation because the identity maps each line onto itself. It remains for the reader to show that the inverse of each dilation is also a dilation. (See Exercise 7, Exercise Set 6–1.) Thus:

The set D of dilations of $\mathbb{P}$, under composition, is a subgroup of the group of transformations of $\mathbb{P}$.

An example of a dilation is the function p that maps each point (x, y) of $\mathbb{P}$ onto the point $(2x + 1, 2y - 1)$. Consider any

206   geometric transformations

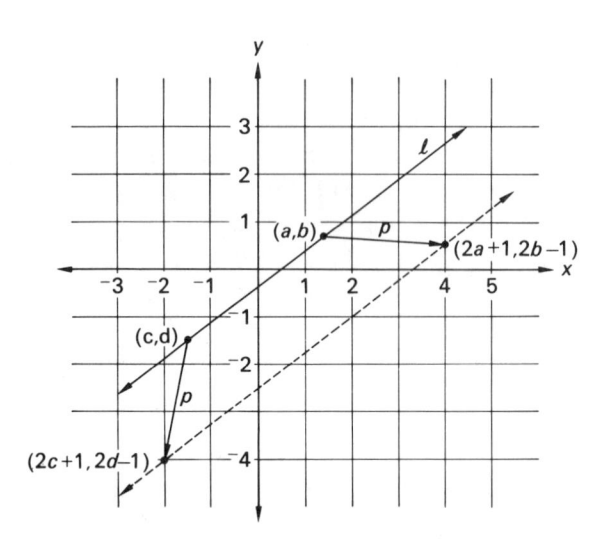

Figure 6–3

line l, and let (a, b) (c, d) be any points on l (see Fig. 6–3); then a direction vector of l is $[c - a, d - b]$. Since

$$(a, b) \xrightarrow{p} (2a + 1, 2b - 1) \quad \text{and} \quad (c, d) \xrightarrow{p} (2c + 1, 2d - 1),$$

a direction vector of the line through the image points is

$$[2c + 1 - 2a - 1, 2d - 1 - 2b + 1] = 2[c - a, d - b].$$

Thus p maps line l onto a line with a parallel direction vector. Moreover, we also see that p maps segments onto segments twice as long; thus p is not an isometry.

In particular, the dilation p defined above maps the triangle with vertices $(1, 0)$, $(0, 1)$, $(0, 0)$ onto the triangle with vertices $(3, {}^-1)$, $(1, 1)$, $(1, {}^-1)$ (Fig. 6–4). The image triangle is similar in shape to the original triangle, with each of its sides parallel to and twice the length of a side of the original. We shall say that

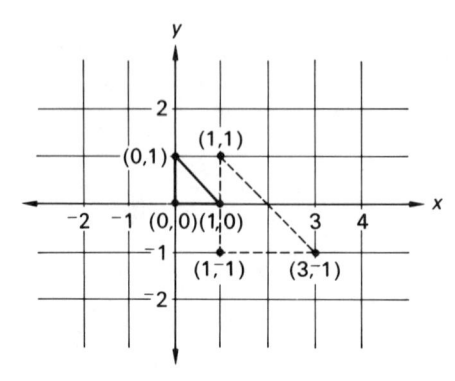

Figure 6–4

dilations map figures onto *similar* figures, although not all transformations yielding similar figures are dilations.

It is instructive to look at the above dilation p as the composite of two functions. If we first map (x, y) onto $(2x, 2y)$ by a function, say m, and then map $(2x, 2y)$ onto $(2x + 1, 2y - 1)$ by a function, say z, we see that $p = z \circ m$. The function m is called a *magnification* about $(0, 0)$, or simply a *magnification;* it has the effect of "stretching" the plane in all directions from the fixed point $(0, 0)$. The function z we recognize as the translation $[1, ^-1]$:

$$(x, y) \xrightarrow{z} (x + 1, y - 1).$$

Using the notation of adding points and multiplying points by numbers, we write

$$(x, y) \xrightarrow{p} 2(x, y) + (1, ^-1)$$

or

$$(x, y) \xrightarrow{m} 2(x, y) \xrightarrow{z} 2(x, y) + (1, ^-1).$$

Note in Fig. 6–5 that m magnifies $\triangle ABO$ to $\triangle CDO$; and then z translates $\triangle CDO$ to $\triangle EFG$. The composite dilation $p = z \circ m$ maps $\triangle ABO$ directly onto $\triangle EFG$.

Figure 6–5

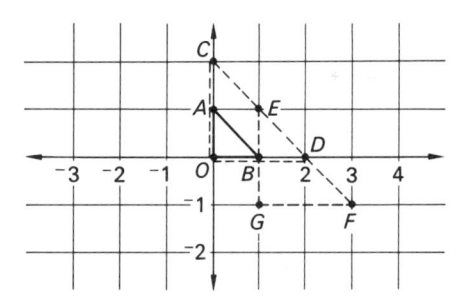

In general, it can be shown that every dilation is of the form of the function p in the above example.

Every dilation p of $\mathbb{P}$ is the composite of a magnification m of $\mathbb{P}$ and a translation z of $\mathbb{P}$: $p = z \circ m$, where $(x, y) \xrightarrow{m} (tx, ty)$ for some nonzero real number t, and $(x, y) \xrightarrow{z} (x + a, y + b)$ for some real numbers a and b. We call t the magnification factor and $[a, b]$ the translation vector. For all (x, y) in $\mathbb{P}$,

$$(x, y) \xrightarrow{p} (tx + a, ty + b).$$

Thus far we have identified two subgroups of the full group T of transformations of $\mathbb{P}$: the subgroup E of isometries (length-preserving) and the subgroup D of dilations (parallel-preserving). Although in the examples we exhibited an isometry that is not a dilation and a dilation that is not an isometry, there *are* functions common to these two subgroups. We need only look among the dilations for functions that preserve lengths of segments. Among these are the translations of $\mathbb{P}$, which we know are isometries and which we recognize as dilations with magnification factor $t = 1$, that is, with the identity as magnification function.

Are there other dilations of $\mathbb{P}$ that are isometries? Consider any two points $P = (x_1, y_1)$, $Q = (x_2, y_2)$ and their images under any dilation

$$d(P) = (tx_1 + a, ty_1 + b), \qquad d(Q) = (tx_2 + a, ty_2 + b).$$

The length of the segment $\overline{PQ}$ is $\sqrt{(x_2 - x_1)^2 + (y_2 - y_1)^2}$, whereas the length of the image segment $\overline{d(P)\,d(Q)}$ is

$$\sqrt{t^2[(x_2 - x_1)^2 + (y_2 - y_1)^2]}.$$

These segments have the same length if and only if $t = 1$ or $t = {}^-1$. We disposed of the case $t = 1$, which yields the subgroup of translations. For the case $t = {}^-1$, we have the set of functions which maps each point (x, y) of $\mathbb{P}$ onto $({}^-x + a, {}^-y + b)$, for any real numbers a and b. Each of these functions is the composite of the fixed function $f: (x, y) \to ({}^-x, {}^-y)$, and some translation $z: [a, b]$.

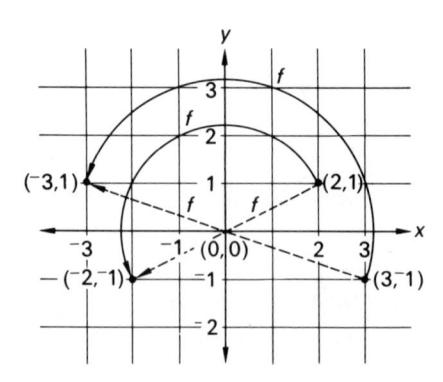

Figure 6–6

If we look at the f-image of several points of $\mathbb{P}$ (Fig. 6–6), we suspect that f is a 180° rotation of the plane about $(0, 0)$. We shall verify this in a later section. Thus a dilation with magnification factor $t = {}^-1$ will be found to be a 180° rotation about a point of $\mathbb{P}$.

To summarize, we found two subgroups of the full group T of (invertible) transformations of $\mathbb{P}$: the isometries E and the dilations D. The intersection of these two subgroups consists of the following: the translations, which are a subgroup of E and a subgroup of D, and the set of 180° rotations about any point. In Exercise Set 6–1, the reader is asked to show that this set of 180° rotations is not a group.

When we study plane geometry, the idea of congruence of figures is defined in terms of isometries of $\mathbb{P}$. We say that figures α and β are congruent if there is a function in the set E of isometries that maps figure α onto figure β. Later the idea of two figures being similar is defined generally in terms of transformations of $\mathbb{P}$ called similarities, of which dilations are special cases. We shall consider these in Section 6–4.

exercise set 6–1

1. The function $(x, y) \xrightarrow{q} (2x, y)$ is not an isometry, but it is invertible. Find q^{-1}; i.e. a function q^{-1} such that $(x, y) \xrightarrow{q} (2x, y) \xrightarrow{q^{-1}} (x, y)$.

2. Given the function $(x, y) \xrightarrow{v} ({}^-3x - 1, {}^-3y + 2)$.
   a) Find the v-image of the triangle with vertices $(1, {}^-1)$, $(3, 2)$, $(4, 1)$.
   b) Show that v is a dilation of $\mathbb{P}$. (Follow the example on page 205 to show that any line l is mapped by v onto a line parallel to l.)
   c) Show that
   $$(x, y) \xrightarrow{w} \left(\frac{{}^-x}{3} - \frac{1}{3}, \frac{{}^-y}{3} + \frac{2}{3} \right)$$
   is the inverse of v.
   d) Verify that w is the inverse of v by finding the w-image of the v-image of the triangle in part (a).
   e) Find the fixed points of v and of w. [For v, find (x, y) such that $({}^-3x - 1, {}^-3y + 2) = (x, y)$; that is, such that ${}^-3x - 1 = x$ and ${}^-3y + 2 = y$.]

3. Given any dilation $(x, y) \xrightarrow{d} (tx + a, ty + b)$, $t \neq 0$.
   a) Show that d has exactly one fixed point
   $$\left(\frac{{}^-a}{t - 1}, \frac{{}^-b}{t - 1} \right) \quad \text{if } t \neq 1.$$

b) Show that d has no fixed points if $t = 1$ and a or b is not zero.
c) Interpret these results in terms of the effects of d for $t \neq 1$ and $t = 1$. (Explain why d may be called a magnification if $t \neq 1$, and a translation if $t = 1$.)

4. Prove: If f and g are functions from set S to S such that $f \circ g = g \circ f = I$ (the identity function), then f and g are 1–1 and onto. [*Hint:* For any b in S, $b = f \circ g(b) = f(g(b)) = f(a)$, showing that f is onto; if $f(a) = f(b)$, then

$$a = g \circ f(a) = g(f(a)) = g(f(b)) = g \circ f(b),$$

and we then have $a = b$, showing that f is 1–1. Fill in details and reasons, and follow similar proof for g.]

*5. Prove: Let G be a group under an operation $*$, and let S be a non-empty subset of the set G. Then S is a subgroup of G if and only if
1) For all a, b in S, $a * b$ is in S.
2) For all a in S, a^{-1} is in S.

6. Apply the theorem proved in Exercise 5 to show that:
a) The set of translations of $\mathbb{P}$ is a subgroup of the group E of isometries of $\mathbb{P}$, under the operation of composition.
b) The set of magnifications of $\mathbb{P}$ [about $(0, 0)$] is a subgroup of the group D of dilations of $\mathbb{P}$, under the operation of composition.

7. Show that the inverse of any dilation is also a dilation. [Let d be a dilation and d^{-1} its inverse. Then d maps $\overline{PQ}$ onto segment $\overline{d(P)\ d(Q)}$ parallel to $\overline{PQ}$. What is the effect of d^{-1} on the segment $\overline{d(P)\ d(Q)}$?]

8. Find the fixed point of the 180° rotation

$$(x, y) \xrightarrow{h} (^-x + a,\ ^-y + b).$$

(See Exercise 3.) Call this the *center* of the 180° rotation. Given any point (c, d), find the function which is a 180° rotation with center (c, d).

9. a) Find the composite $h_2 \circ h_1$, where $(x, y) \xrightarrow{h_1} (^-x + 2,\ ^-y + 1)$, $(x, y) \xrightarrow{h_2} (^-x + 3,\ ^-y - 2)$ are 180° rotations. Is $h_2 \circ h_1$ a 180° rotation?
b) Explain why the results of part (a) show that the set of 180° rotations is not a group.

10. Continue Exercise 9(a) to show that the composite of any two 180° rotations is a translation. Are $h_1 \circ h_2$ and $h_2 \circ h_1$ the same translation?

11. Continue Exercises 9 and 10 to show that the intersection of the subgroup E of isometries and the subgroup D of dilations is itself a subgroup (of the group of transformations). In fact, show that generally the intersection of two subgroups of a group is a subgroup of the group.

6–3
orthogonal transformations and linear functions

In Chapter 4 it was decided that isometries (rigid motions) of $\mathbb{P}$ can be classified in terms of their fixed points. For example, an isometry with exactly one fixed point is a rotation about that point; an isometry with every point of a line a fixed point (and only these fixed points) is a reflection in that line. If an isometry has two distinct fixed points, then every point on the line determined by these is also a fixed point (as you will see in Example 1 below), and the isometry is the identity function or is again a reflection. Furthermore, if an isometry has three or more noncollinear fixed points, it is the identity function. (Why?) Finally, if an isometry has no fixed points, then it is a translation or a glide reflection.

Our plan for studying the group of isometries is to consider first the set of those isometries that have the origin $(0, 0)$ as a fixed point. (This excludes the translations immediately.) Let us call these the *orthogonal transformations* of $\mathbb{P}$. Then each isometry of $\mathbb{P}$ will be the composite of an orthogonal transformation [a rotation about $(0, 0)$ or a reflection in a line through $(0, 0)$] and a translation. Let us look at several examples of orthogonal transformations.

• *Example 1.* $(x, y) \xrightarrow{f} (x, \bar{y})$.

This function f has $(0, 0)$ as a fixed point, and we see that it is an isometry by noting that it maps $P = (x_1, y_1)$, $Q = (x_2, y_2)$ onto $f(P) = (x_1, \bar{y}_1)$, $f(Q) = (x_2, \bar{y}_2)$, so that the length of $\overline{PQ}$ is $\sqrt{(x_2 - x_1)^2 + (y_2 - y_1)^2}$ and the length of the image segment $\overline{f(P)f(Q)}$ is $\sqrt{(x_2 - x_1)^2 + (\bar{y}_2 + y_1)^2}$; that is, the length of the image segment is the same as the length of the segment. After we have drawn figures of a few points and their images (see Fig. 6–7), we suspect that f is the reflection of $\mathbb{P}$ in the x-axis. This suspicion is justified when we find that every point on the x-axis is a fixed point of f.

212   geometric transformations

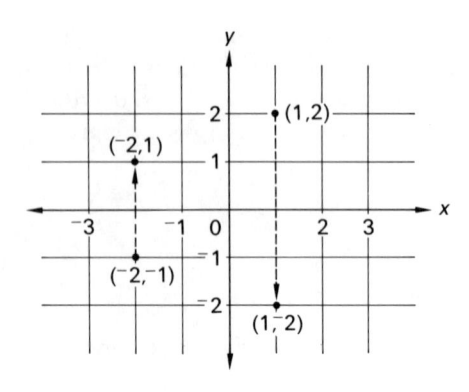

Figure 6–7

• *Example 2.* $(x, y) \xrightarrow{q} (^-y, x)$.

Again $(0, 0)$ is a fixed point, and again we find that the length of the segment $\overline{PQ}$ from $P = (x_1, y_1)$ to $Q = (x_2, y_2)$ is the same as the length of the image segment $\overline{g(P)\,g(Q)}$. Thus g is an isometry with $(0, 0)$ fixed. A few points and their images (Fig. 6–8) suggest that g is a 90° (counterclockwise) rotation of $\mathbb{P}$ about $(0, 0)$. The fact that $(0, 0)$ is the *only* fixed point guarantees that g is a rotation about $(0, 0)$.

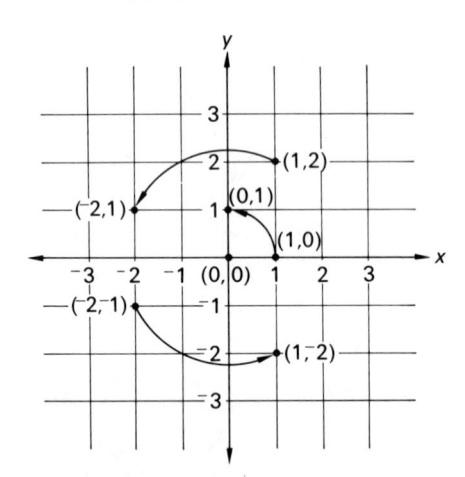

Figure 6–8

• *Example 3.* $(x, y) \xrightarrow{h} (\tfrac{3}{5}x - \tfrac{4}{5}y, \tfrac{4}{5}x + \tfrac{3}{5}y)$.

A short computation shows that $(0, 0)$ is a fixed point, and the only fixed point, of h. Let

$$\tfrac{3}{5}x - \tfrac{4}{5}y = x, \qquad \tfrac{4}{5}x + \tfrac{3}{5}y = y;$$

then

$$-\tfrac{2}{5}x - \tfrac{4}{5}y = 0, \qquad \tfrac{4}{5}x - \tfrac{2}{5}y = 0,$$

so that $$x + 2y = 0, \quad 2x - y = 0,$$
which implies that $x = 0$ and $y = 0$.

A somewhat longer computation shows that h also preserves lengths of segments: $P = (x_1, y_1)$, $Q = (x_2, y_2)$; then
$$PQ = \sqrt{(x_2 - x_1)^2 + (y_2 - y_1)^2}$$
and

$h(P)h(Q)$
$$= \sqrt{(\tfrac{3}{5}(x_2 - x_1) - \tfrac{4}{5}(y_2 - y_1))^2 + (\tfrac{4}{5}(x_2 - x_1) + \tfrac{3}{5}(y_2 - y_1))^2}$$
$$= \sqrt{\tfrac{9}{25}(x_2 - x_1)^2 - \tfrac{24}{25}(x_2 - x_1)(y_2 - y_1) + \tfrac{16}{25}(y_2 - y_1)^2}$$
$$\overline{ + \tfrac{16}{25}(x_2 - x_1)^2 + \tfrac{24}{25}(x_2 - x_1)(y_2 - y_1) + \tfrac{9}{25}(y_2 - y_1)^2}$$
$$= \sqrt{(x_2 - x_1)^2 + (y_2 - y_1)^2}.$$

Thus h is an orthogonal transformation. The fact that h has $(0, 0)$ as its *only* fixed point then tells us that h is a rotation about $(0, 0)$ (Fig. 6–9). The angle θ of rotation is indicated by the fact that $(1, 0) \xrightarrow{h} (\tfrac{3}{5}, \tfrac{4}{5})$.

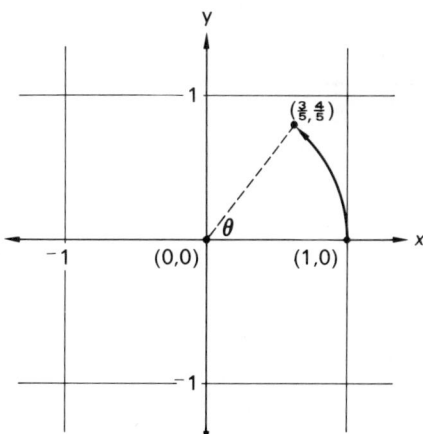

Figure 6–9

linear functions The three examples of orthogonal transformations given above all followed the same forms of representation. Functions of $\mathbb{P}$ of the form
$$(x, y) \rightarrow (ax + by, cx + dy)$$
are called *linear* functions.

In Example 1, $a = 1$, $b = 0$, $c = 0$, $d = {}^-1$.

In Example 2, $a = 0$, $b = {}^-1$, $c = 1$, $d = 0$.
In Example 3, $a = \frac{3}{5}$, $b = {}^-\frac{4}{5}$, $c = \frac{4}{5}$, $d = \frac{3}{5}$.

We note immediately that linear functions have $(0, 0)$ as a fixed point. (See Exercise 6 of Exercise Set 6–2.) It can also be shown that every orthogonal transformation is a linear function. Our problem now is to determine which linear functions are orthogonal.

First we observe that some linear functions are not orthogonal; for example, the function

$$(x, y) \xrightarrow{k} (x + y, 2x + 2y)$$

is linear, with $a = 1, b = 1, c = 2, d = 2$, but it is not orthogonal. It is not even a (invertible) transformation: Since k maps $(2, {}^-1)$, $(3, {}^-2)$, and other points onto the same point $(1, 2)$, we see that k is not 1–1 and hence not invertible (Fig. 6–10). In fact, k maps the whole plane $\mathbb{P}$ onto the line $\{(x, y) \mid 2x - y = 0\}$.

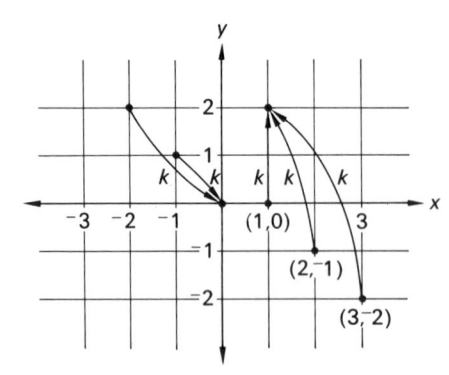

Figure 6–10

Let us attack the problem directly. Consider any linear function

$$(x, y) \xrightarrow{L} (ax + by, cx + dy)$$

and any two distinct points $P = (x_1, y_1)$, $Q = (x_2, y_2)$. By direct computation, let us determine the relations between the numbers a, b, c, d for which the segment $\overline{PQ}$ has the same length as its image segment $\overline{L(P) L(Q)}$.

$$L(P) = (ax_1 + by_1, cx_1 + dy_1),$$
$$L(Q) = (ax_2 + by_2, cx_2 + dy_2),$$

$$L(P)L(Q)$$
$$= \sqrt{(a(x_2 - x_1) + b(y_2 - y_1))^2 + (c(x_2 - x_1) + d(y_2 - y_1))^2}$$
$$= \sqrt{a^2(x_2 - x_1)^2 + 2ab(x_2 - x_1)(y_2 - y_1) + b^2(y_2 - y_1)^2}$$
$$\overline{+ c^2(x_2 - x_1)^2 + 2cd(x_2 - x_1)(y_2 - y_1) + d^2(y - y_1)^2}$$
$$= \sqrt{(a^2 + c^2)(x_2 - x_1)^2 + 2(ab + cd)(x_2 - x_1)(y_2 - y_1)}$$
$$\overline{+ (b^2 + d^2)(y_2 - y_1)^2}.$$

Compare the length of the image segment $\overline{L(Q)L(P)}$ with $PQ = \sqrt{(x_2 - x_1)^2 + (y_2 - y_1)^2}$. We see that if $L(P)L(Q) = PQ$ for any choice of (x_1, y_1), (x_2, y_2), we must have

(*) $\qquad a^2 + c^2 = 1, \qquad ab + cd = 0, \qquad b^2 + d^2 = 1.$

Thus we have found the linear functions that are orthogonal: those for which the relations (*) hold.

In Example 1 above, we had $a = 1$, $b = 0$, $c = 0$, $d = {}^-1$, so that $a^2 + c^2 = 1$, $ab + cd = 0$, $b^2 + d^2 = 1$. The reader should check that Examples 2 and 3 also satisfy the relations (*).

Let us attempt to characterize all linear functions that are orthogonal transformations by determining a, b, c, d such that the relations (*) hold. First we note that a, b cannot both be zero and c, d cannot both be zero; otherwise L would not be 1–1, that is, not invertible. Second, consider these cases.

Case 1. If $a = 0$, then $c^2 = 1$. Then the condition $ab + cd = 0$ requires that $d = 0$. Thus $b^2 = 1$ and $c^2 = 1$. With $b = 1$ or $b = {}^-1$, and $c = 1$ or $c = {}^-1$, and $a = 0$, $d = 0$ we have four possibilities for L:

$$(x, y) \to ({}^-y, x); \qquad (x, y) \to ({}^-y, {}^-x);$$
$$(x, y) \to (y, x); \qquad (x, y) \to (y, {}^-x).$$

Case 2. If $a \neq 0$, then $d = \left(\dfrac{d}{a}\right) a$, and the condition

$$ab + cd = 0 \qquad \text{implies that} \qquad ab + c\left(\dfrac{d}{a}\right) a = 0,$$

which implies that

$$b = -\left(\dfrac{d}{a}\right) c.$$

Substituting these results into $b^2 + d^2 = 1$ gives
$$\left(\frac{d}{a}\right)^2 c^2 + \left(\frac{d}{a}\right)^2 a^2 = 1,$$
which is true if and only if
$$\left(\frac{d}{a}\right)^2 (c^2 + a^2) = 1.$$
But $c^2 + a^2 = 1$, which implies that $\left(\frac{d}{a}\right)^2 = 1$. This gives us the characterization we desire. For any a, c such that $a^2 + c^2 = 1$, we chose $b = -\left(\frac{d}{a}\right)c$, $d = \left(\frac{d}{a}\right)a$, where $\frac{d}{a} = 1$ or $\frac{d}{a} = {}^-1$. Note that these results also include Case 1.

We have proved the following theorem.

▶ *Theorem 2.* The linear function
$$(x, y) \xrightarrow{L} (ax + by, cx + dy)$$
is an orthogonal transformation if and only if a, c are any numbers such that $a^2 + c^2 = 1$, and $b = {}^-c$, $d = a$ or $b = c$, $d = {}^-a$. Thus the orthogonal transformations of $\mathbb{P}$ are of two forms:

1) $(x, y) \to (ax - cy, cx + ay)$,
2) $(x, y) \to (ax + cy, cx - ay)$,

where $a^2 + c^2 = 1$.

At this point a reader who has studied trigonometry will probably cry "Eureka" and quickly identify all pairs of numbers a, c such that $a^2 + c^2 = 1$. For those whose knowledge of trigo-

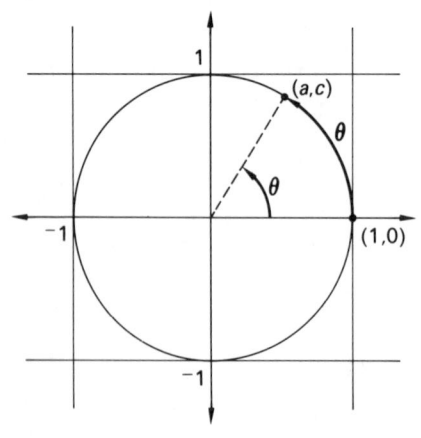

Figure 6–11

nometry is limited, we recall that for any angle (or arc) θ the trigonometric functions cos and sin are defined as $\cos \theta = a$, $\sin \theta = c$, where $a^2 + c^2 = 1$; that is, (a, c) is a point on the circle with center $(0, 0)$ and radius 1, and θ is the length of the arc from $(1, 0)$ counterclockwise to (a, c) on the circle (Fig. 6–11).

Now we have an alternate way of writing an orthogonal transformation: For any angle θ from 0 to 2π (from 0° to 360°),

1) $(x, y) \rightarrow ((\cos \theta)x - (\sin \theta)y, (\sin \theta)x + (\cos \theta)y)$,

2) $(x, y) \rightarrow ((\cos \theta)x + (\sin \theta)y, (\sin \theta)x - (\cos \theta)y)$.

We can show that transformations of type (1) have only $(0, 0)$ as a fixed point and that transformations of type (2) have all points on the line $cx = (a + 1)y$ fixed (see Exercise 10 of Exercise Set 6–2). More difficult computations supply the angles of rotation and angles to lines of reflections. These computations show that orthogonal transformations of type (1) are rotations of $\mathbb{P}$ about $(0, 0)$ counterclockwise through angle θ; those of type (2) are reflections of $\mathbb{P}$ in a line through $(0, 0)$ inclined at an angle of $\frac{\theta}{2}$ with the positive x-axis.

Again consider the three examples at the beginning of this section. In Example 1, $a = 1$ and $c = 0$ imply that $\cos \theta = 1$ and $\sin \theta = 0$. Thus $\theta = 0$. Since $d = {}^-a$, the function is of type (2); that is, a reflection in the line through $(0, 0)$ inclined at a 0-angle to the x-axis, namely the reflection in the x-axis. In Example 2, $a = 0$ and $c = 1$ imply that $\cos \theta = 0$ and $\sin \theta = 1$; thus $\theta = \frac{\pi}{2}$ (or 90°). Since $d = a$, the function is a rotation of $\mathbb{P}$ about $(0, 0)$ through an angle of $\frac{\pi}{2}$ (90°), that is, a 90° rotation. In Example 3, $a = \frac{3}{5}$ and $b = \frac{4}{5}$ imply that $\cos \theta = \frac{3}{5}$, $\sin \theta = \frac{4}{5}$, so that $\theta \approx 53°8'$. Since $a = d$, this function is a rotation about $(0, 0)$ through an angle of about 53°8'.

As another example, consider the function

$$(x, y) \xrightarrow{k} ({}^-y, {}^-x).$$

In this case, $a = 0$, $b = {}^-1$, $c = {}^-1$, $d = 0$, so that the function k is an orthogonal transformation of type (2). Since $\cos \theta = 0$,

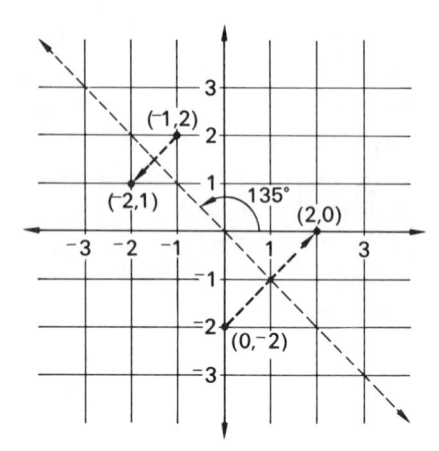

Figure 6–12

$\sin \theta = {}^-1$ imply that $\theta = \dfrac{3\pi}{2}$ (270°), the function is a reflection in a line through (0, 0) inclined at an angle $\dfrac{\theta}{2} = \dfrac{3\pi}{4}$ (135°) to the positive x-axis (Fig. 6–12). As a check on this result, we can compute the fixed points of the function k; $({}^-y, {}^-x) = (x, y)$ implies ${}^-y = x$ and ${}^-x = y$; that is, $y = {}^-x$. Thus, all points on the line $y = {}^-x$ are fixed.

exercise set 6–2

1. Show that if f is an isometry and if f has two distinct fixed points, P and Q, then every point on $\overleftrightarrow{PQ}$ is a fixed point of f (Fig. 6–13). (Let V be any other point on $\overleftrightarrow{PQ}$ and assume that V is not a fixed point of f. Then $f(P) = P$, $f(Q) = Q$, but $f(V) \neq V$. Let $f(V) = W$. Now what can be said of the length of segments $\overline{PV}$, $\overline{VQ}$, and $\overline{PQ}$ versus the length of segments $\overline{f(P)f(V)}$, $\overline{f(V)f(Q)}$, and $\overline{f(P)f(Q)}$? Remember that f is an isometry!)

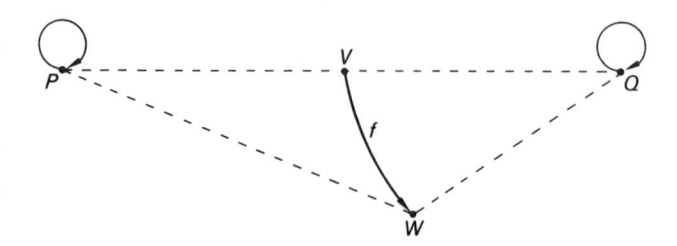

Figure 6–13

2. Show that if f is an isometry and if f has at least three noncollinear fixed points, then every point of $\mathbb{P}$ is a fixed point of f (that is, f is the identity function). (Choose two of the fixed points, P and Q. What does Exercise 1 say about all the points on line $\overleftrightarrow{PQ}$? There is a third fixed point R, not on $\overleftrightarrow{PQ}$. What does Exercise 1 say about every point on every line through R and a point of $\overleftrightarrow{PQ}$?)

3. For each of the following isometries of $\mathbb{P}$, determine all its fixed points, and on the basis of this information classify the isometry.
   a) $(x, y) \xrightarrow{a} (x, {}^{-}y)$
   b) $(x, y) \xrightarrow{b} ({}^{-}y, x)$
   c) $(x, y) \xrightarrow{c} ({}^{-}y, {}^{-}x)$
   d) $(x, y) \xrightarrow{d} (x - 3, y + 2)$
   e) $(x, y) \xrightarrow{e} ({}^{-}x - 2, {}^{-}y + 1)$
   f) $(x, y) \xrightarrow{f} ({}^{-}x + 1, y - 2)$
   g) $(x, y) \xrightarrow{g} (\tfrac{5}{13}x + \tfrac{12}{13}y, \tfrac{12}{13}x - \tfrac{5}{13}y)$
   h) $(x, y) \xrightarrow{h} (x - 1, {}^{-}y - 2)$

4. In Exercise 3, decide which isometries are orthogonal transformations.

5. For each of the following functions of $\mathbb{P}$, decide (i) which are linear, (ii) which are isometries, (iii) which are orthogonal transformations, and (iv) classify the isometries by their fixed points.
   a) $(x, y) \xrightarrow{a} (2x, y)$
   b) $(x, y) \xrightarrow{b} ({}^{-}x + 1, {}^{-}y - 1)$
   c) $(x, y) \xrightarrow{c} (x + y, x - y)$
   d) $(x, y) \xrightarrow{d} (\tfrac{4}{5}x + \tfrac{3}{5}y, \tfrac{3}{5}x - \tfrac{4}{5}y)$
   e) $(x, y) \xrightarrow{e} (xy, x)$
   f) $(x, y) \xrightarrow{f} (\tfrac{12}{13}x - \tfrac{5}{13}y, \tfrac{5}{13}x + \tfrac{12}{13}y)$

6. Show that every linear function of $\mathbb{P}$ has $(0, 0)$ as a fixed point. (Show that the equations $ax + by = x$ and $cx + dy = y$ are true for $x = 0, y = 0$ regardless of the values of a, b, c, d.)

7. Show that each linear function L of $\mathbb{P}$ maps each line through $(0, 0)$ onto a line through $(0, 0)$, provided L is not the function $(x, y) \to (0, 0)$ (Fig. 6–14). (Consider any line k through $(0, 0)$ and (r, s) and let L be any nonzero linear function $(x, y) \xrightarrow{L} (ax + by, cx + dy)$. Then $k = \{(x, y) \mid sx - ry = 0\}$. Since $(0, 0) \xrightarrow{L} (0, 0)$ and $(r, s) \xrightarrow{L} (ar + bs, cr + ds) = P$, the line l

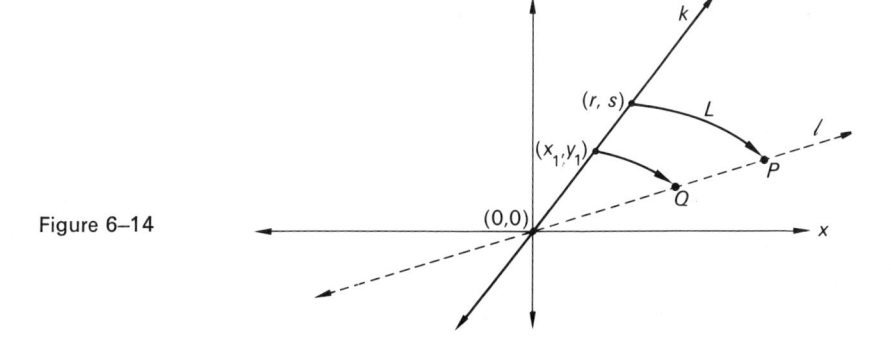

Figure 6–14

through $(0, 0)$ and P is $\{(x, y) \mid (cr + ds)x - (ar + bs)y = 0\}$. Now show that any point (x_1, y_1) of k is mapped by linear function L onto a point of l. That is, $(x_1, y_1) \xrightarrow{L} (ax_1 + by_1, cx_1 + dy_1) = Q$; and $sx_1 - ry_1 = 0$; now show that Q satisfies the equation of line l.)

8. Show that if $a = 0$ and $b = 0$, the linear function

$$(x, y) \xrightarrow{L} (ax + by, cx + dy)$$

is not 1–1, and hence is not invertible.

9. If L is an isometry, then

$$(x, y) \xrightarrow{L} (ax - cy, cx + ay), \qquad a^2 + c^2 = 1,$$

or

$$(x, y) \xrightarrow{L} (ax + cy, cx - ay), \qquad a^2 + c^2 = 1.$$

On page 215 are listed the four isometries obtained when $a = 0$. Find the four isometries for $c = 0$. For this set S of eight orthogonal transformations:
  a) Classify each as a rotation or reflection and give its angle of rotation or the angle of its line of reflection.
  b) Show that each is a symmetry of the square with vertices at $(1, 0)$, $(0, 1)$, $(^-1, 0)$, $(0, ^-1)$. (Compare with Exercise 7, Exercise Set 4–3.)

*10. For those readers with some background in algebra and trigonometry:
  a) Show that except for the identity function there is one and only one fixed point of the orthogonal transformation of form (1), namely $(0, 0)$, and hence the transformation is a rotation about $(0, 0)$.
  b) Show that an orthogonal transformation of form (1) other than the identity maps $(1, 0)$ onto $(\cos \theta, \sin \theta)$, and hence it is a rotation about $(0, 0)$ through angle θ.
  c) Show that an orthogonal transformation of form (2) has each point of the line $\{(x, y) \mid cx - (a + 1)y = 0, a^2 + c^2 = 1\}$ a fixed point, and hence is a reflection in that line.
  d) Since $a = \cos \theta$, $c = \sin \theta$, show that the angle ϕ of the line of reflection of form (2) is such that

$$\tan \phi = \frac{\sin \theta}{\cos \theta + 1}.$$

Then show that $\phi = \dfrac{\theta}{2}$.

6–4
matrices and composition of linear functions

Our plan is to describe any isometry of $\mathbb{P}$ as a composite of an orthogonal transformation and a translation. It was an easy matter in Chapter 5 to compose translations; this led to vector addition. Before going on to the more general problem of composing two isometries, we need techniques for composing orthogonal transformations.

Let us attack this problem by taking on the broader problem of composing linear functions, since both orthogonal transformations and magnifications are linear functions. We begin by writing any linear function L in the form

$$L: \ (x, y) \to (ax + by, cx + dy)$$

and developing a simpler notation for this representation. We shall agree to write $\begin{pmatrix} x \\ y \end{pmatrix}$ for the point (x, y). Then we can write

$$L: \ \begin{pmatrix} x \\ y \end{pmatrix} \to \begin{pmatrix} ax + by \\ cx + dy \end{pmatrix}.$$

Since we always keep the same order, with x first and y second, we can denote L by the square array $\begin{pmatrix} a & b \\ c & d \end{pmatrix}$. The notation is then:

$$\underset{\substack{\uparrow \\ \text{Linear} \\ \text{function } L}}{\begin{pmatrix} a & b \\ c & d \end{pmatrix}} \ \underset{\text{maps}}{} \ \underset{\substack{\uparrow \\ \text{Point} \\ (x, y)}}{\begin{pmatrix} x \\ y \end{pmatrix}} \ \underset{\text{onto}}{=} \ \underset{\substack{\uparrow \\ \text{Point} \\ (ax + by, cx + dy)}}{\begin{pmatrix} ax + by \\ cx + dy \end{pmatrix}}$$

We say that L is represented by the *matrix* $\begin{pmatrix} a & b \\ c & d \end{pmatrix}$, where a, b, c, d are real numbers. $\begin{pmatrix} a & b \\ c & d \end{pmatrix}$ is called a *two-by-two matrix* because it has two rows and two columns.

Again going back to two of the examples of Section 6–3, pages 211–213, we have:

- *Example 1.* The reflection $(x, y) \xrightarrow{f} (x, {}^-y)$ is represented by the matrix $\begin{pmatrix} 1 & 0 \\ 0 & -1 \end{pmatrix}$, since $(x, ^-y) = (1x + 0y, 0x - 1y)$. Then:

$$\begin{pmatrix} 1 & 0 \\ 0 & -1 \end{pmatrix}\begin{pmatrix} x \\ y \end{pmatrix} = \begin{pmatrix} 1x + 0y \\ 0x - 1y \end{pmatrix} = \begin{pmatrix} x \\ -y \end{pmatrix}.$$

The computation

$$\begin{pmatrix} 1 & 0 \\ 0 & -1 \end{pmatrix}\begin{pmatrix} 2 \\ 3 \end{pmatrix} = \begin{pmatrix} 1 \cdot 2 + 0 \cdot 3 \\ 0 \cdot 2 - 1 \cdot 3 \end{pmatrix}$$

indicates that f maps point $(2, 3)$ onto $(2, ^-3)$.

- *Example* 3. The rotation h is represented by the matrix

$$\begin{pmatrix} \frac{3}{5} & -\frac{4}{5} \\ \frac{4}{5} & \frac{3}{5} \end{pmatrix},$$

so that

$$\begin{pmatrix} \frac{3}{5} & -\frac{4}{5} \\ \frac{4}{5} & \frac{3}{5} \end{pmatrix}\begin{pmatrix} x \\ y \end{pmatrix} = \begin{pmatrix} \frac{3}{5}x - \frac{4}{5}y \\ \frac{4}{5}x + \frac{3}{5}y \end{pmatrix}.$$

For example, the computation

$$\begin{pmatrix} \frac{3}{5} & -\frac{4}{5} \\ \frac{4}{5} & \frac{3}{5} \end{pmatrix}\begin{pmatrix} 5 \\ -10 \end{pmatrix} = \begin{pmatrix} \frac{3}{5} \cdot 5 - \frac{4}{5}(^-10) \\ \frac{4}{5} \cdot 5 - \frac{3}{5} \cdot 10 \end{pmatrix} = \begin{pmatrix} 11 \\ -2 \end{pmatrix}$$

shows that h maps the point $(5, ^-10)$ onto $(11, ^-2)$.

With matrix notation for linear functions, we are able to streamline the computation of composite functions. For example, consider the linear function

$$R: \begin{array}{l} x \to x - y \\ y \to 2x \end{array}$$

which maps the point $(^-1, 1)$ onto $(^-2, ^-2)$. Then choose another linear function, say

$$S: \begin{array}{l} x \to ^-x \\ y \to x + y. \end{array}$$

This function maps point $(^-2, ^-2)$ onto $(2, ^-4)$. Thus the function R followed by the function S has the effect of mapping point $(^-1, 1)$ onto $(2, ^-4)$. In general,

$$\begin{pmatrix} x \\ y \end{pmatrix} \xrightarrow{R} \begin{pmatrix} x - y \\ 2x \end{pmatrix} \xrightarrow{S} \begin{pmatrix} ^-(x - y) \\ (x - y) + (2x) \end{pmatrix},$$

so that
$$\begin{pmatrix}x\\y\end{pmatrix} \xrightarrow{SR} \begin{pmatrix}-x+y\\3x-y\end{pmatrix}.$$

Let us retrace the steps of this example, using matrix notation.
$$R: \begin{pmatrix}1 & -1\\2 & 0\end{pmatrix}, \quad S: \begin{pmatrix}-1 & 0\\1 & 1\end{pmatrix}$$

$$\underbrace{\begin{pmatrix}-1 & 0\\1 & 1\end{pmatrix}}_{S}\underbrace{\left[\underbrace{\begin{pmatrix}1 & -1\\2 & 0\end{pmatrix}}_{\text{Image of }(x,y)\text{ under }R}\underbrace{\begin{pmatrix}x\\y\end{pmatrix}}_{}\right]}_{} = \underbrace{\begin{pmatrix}-1 & 0\\1 & 1\end{pmatrix}}_{S}\underbrace{\begin{pmatrix}x-y\\2x\end{pmatrix}}_{R(x,y)}$$

$$= \underbrace{\begin{pmatrix}-(x-y)+0\cdot 2x\\1\cdot(x-y)+1\cdot 2x\end{pmatrix}}_{S(R(x,y))} = \underbrace{\begin{pmatrix}-x+y\\3x-y\end{pmatrix}}_{SR(x,y)}$$

$$= \underbrace{\begin{pmatrix}-1 & 1\\3 & -1\end{pmatrix}}_{SR}\begin{pmatrix}x\\y\end{pmatrix}$$

We write the composite of R followed by S as $S \circ R$ or simply as SR to show that R operates first on a point (x, y) and then S operates on the R-image of (x, y). The above example shows that the composite SR is represented by the matrix

$$\begin{pmatrix}-1 & 1\\3 & -1\end{pmatrix}.$$

We call the matrix of SR the *product* of the matrices of S and R, in that order:

$$-1\cdot 1 + 0\cdot 2 = -1$$

$$\underbrace{\begin{pmatrix}-1 & 0\\1 & 1\end{pmatrix}}_{S}\underbrace{\begin{pmatrix}1 & -1\\2 & 0\end{pmatrix}}_{R} = \underbrace{\begin{pmatrix}-1 & 1\\3 & -1\end{pmatrix}}_{SR}.$$

The example above shows how to compute products of matrices. For example, to find the entry of SR in the first row, first column, we take the first row of S and the first column of R, multiply corresponding entries, and add. To find the entry of SR in the first row, second column, we take the first row of S and the second column of R, multiply corresponding entries and add. And so on.

As another example, consider the linear functions $(x, y) \xrightarrow{U} (y, 2x + y)$ and $(x, y) \xrightarrow{V} (x, x + 2y)$. The composite function VU is found as follows: The matrix of V is

$$\begin{pmatrix} 1 & 0 \\ 1 & 2 \end{pmatrix} \quad \text{and of } U \text{ is} \quad \begin{pmatrix} 0 & 1 \\ 2 & 1 \end{pmatrix}.$$

Then the matrix of VU is

$$\begin{pmatrix} 1 & 0 \\ 1 & 2 \end{pmatrix} \begin{pmatrix} 0 & 1 \\ 2 & 1 \end{pmatrix} = \begin{pmatrix} 0 & 1 \\ 4 & 3 \end{pmatrix}.$$

Thus

$$(x, y) \xrightarrow{VU} (y, 4x + 3y).$$

It is instructive to compute the composite UV:

$$\begin{pmatrix} 0 & 1 \\ 2 & 1 \end{pmatrix} \begin{pmatrix} 1 & 0 \\ 1 & 2 \end{pmatrix} = \begin{pmatrix} 1 & 2 \\ 3 & 2 \end{pmatrix},$$

$$(x, y) \xrightarrow{UV} (x + 2y, 3x + 2y).$$

This shows that composition of linear functions is *not* a commutative operation, because we see that matrix multiplication is not commutative. The composite of two linear functions is represented by the product of their matrices, and we know that composition of functions is an associative operation. Hence, multiplication of matrices is an associative operation.

Let us examine the set L of all linear functions of $\mathbb{P}$. Certainly the composite of two linear functions is a linear function, and composition of functions is generally associative. The identity function is also linear:

$$(x, y) \xrightarrow{I} (x, y),$$

where the matrix of I is

$$\begin{pmatrix} 1 & 0 \\ 0 & 1 \end{pmatrix}.$$

If every linear function had an inverse, we could conclude that the set L of linear functions would be a subgroup of the group of transformations of $\mathbb{P}$.

We can quickly determine whether every linear function has an inverse by looking at some examples. In Exercise 8, Exercise Set 6–2, the linear function $(x, y) \xrightarrow{L} (0, x + y)$ was shown to be not 1–1, and hence not invertible. In order to determine which linear functions *do* have inverses (if any), we must answer this question:

For any given matrix $\begin{pmatrix} a & b \\ c & d \end{pmatrix}$, can we find a matrix $\begin{pmatrix} u & v \\ r & s \end{pmatrix}$ such that

(1) $$\begin{pmatrix} a & b \\ c & d \end{pmatrix} \begin{pmatrix} u & v \\ r & s \end{pmatrix} = \begin{pmatrix} 1 & 0 \\ 0 & 1 \end{pmatrix}?$$

If such a matrix $\begin{pmatrix} u & v \\ r & s \end{pmatrix}$ exists, then upon multiplying matrices we would have

(2) $$\begin{pmatrix} au + br & av + bs \\ cu + dr & cv + ds \end{pmatrix} = \begin{pmatrix} 1 & 0 \\ 0 & 1 \end{pmatrix}.$$

Two matrices are equal if and only if they represent the same linear function, that is, if and only if their corresponding entries are equal. Thus when we equate entries of the equal matrices in (2), we have four equations in which a, b, c, d are fixed and u, v, r, s are to be found, if possible:

$$au + br = 1 \qquad av + bs = 0$$
$$cu + dr = 0 \qquad cv + ds = 1.$$

Solving for u and r in the first pair of equations (see Exercise 7, Exercise Set 6–3), we obtain

$$u = \frac{d}{ad - bc}, \qquad r = \frac{-c}{ad - bc},$$

and solving for v and s in the second pair, we have

$$v = \frac{-b}{ad - bc}, \qquad s = \frac{a}{ad - bc}.$$

We conclude that there is one matrix $\begin{pmatrix} u & v \\ r & s \end{pmatrix}$ that satisfies equation (1) if and only if $ad - bc \neq 0$ (u, v, r, s would not be defined if $ad - bc = 0$). We state these results as a theorem.

▶ *Theorem 3.* The linear function with matrix $\begin{pmatrix} a & b \\ c & d \end{pmatrix}$ has an inverse if and only if $ad - bc \neq 0$. The inverse, when it exists, has matrix

$$\begin{pmatrix} \dfrac{d}{ad-bc} & \dfrac{-b}{ad-bc} \\[2mm] \dfrac{-c}{ad-bc} & \dfrac{a}{ad-bc} \end{pmatrix}.$$

Our question has been answered. Some linear functions do not have inverses and therefore the set of all linear functions is not a group. Note that the noninvertible linear function

$$(x, y) \to (0, x + y)$$

has $a = 0$, $b = 0$, $c = 1$, $d = 1$, so that $ad - bc = 0$, in agreement with Theorem 3.

Let us rephrase the question. Does the set of all *invertible* linear functions form a group under composition? We must decide only whether the composite of two invertible functions is invertible. The answer is yes. (See Exercise 8, Exercise Set 6–3.)

We can finally classify the orthogonal transformations and the magnifications with center $(0, 0)$ as subgroups of the group of all invertible linear functions, which in turn is a subgroup of all (invertible) transformations. This classification is complete as soon as we verify that all orthogonal transformations and magnifications are invertible. Let $\begin{pmatrix} a & -c \\ c & a \end{pmatrix}$ be the matrix of any rotation about $(0, 0)$ with $a^2 + c^2 = 1$. Then $b = -c$, $d = a$, and $ad - bc = a^2 + c^2 = 1 \neq 0$ for any a, c. Thus every rotation about $(0, 0)$ is invertible. (What is the inverse of $\begin{pmatrix} a & -c \\ c & a \end{pmatrix}$? Does it also represent a rotation?) Let $\begin{pmatrix} a & c \\ c & -a \end{pmatrix}$ be the matrix of any reflection in a line through $(0, 0)$, with $a^2 + c^2 = 1$. Then $b = c$, $d = -a$, and $ad - bc = -a^2 - c^2 = -1 \neq 0$. Thus every reflection in a line through $(0, 0)$ is invertible. (Find the inverse matrix; is it the matrix of a reflection?) Finally, let $\begin{pmatrix} t & 0 \\ 0 & t \end{pmatrix}$ be the matrix of any magnification $t \neq 0$. Then $a = t$, $b = 0$, $c = 0$, $d = t$, and $ad - bc = t^2$, which is never zero if $t \neq 0$. Hence every magnification has an inverse. (What is the inverse matrix; is it the matrix of a magnification?)

The number $ad - bc$ associated with the matrix $\begin{pmatrix} a & b \\ c & d \end{pmatrix}$ is called the *determinant* of the linear function associated with $\begin{pmatrix} a & b \\ c & d \end{pmatrix}$ because it determines whether or not the linear function has an inverse.

▶ *Definition:* If the matrix of L is $\begin{pmatrix} a & b \\ c & d \end{pmatrix}$, the number $ad - bc$ is called the *determinant* of L, written det L.

Thus we have proved the following theorem.

▶ *Theorem 4.* For any linear function L from $\mathbb{P}$ to $\mathbb{P}$, det $L \neq 0$ if and only if L is invertible.

Note that the determinant of an orthogonal transformation is 1 or ⁻1 according as the function is a rotation or a reflection. The determinant of a magnification with magnification factor t is t^2.

exercise set 6–3

1. The matrix of an orthogonal transformation is called an orthogonal matrix. What are the relations among a, b, c, d such that $\begin{pmatrix} a & b \\ c & d \end{pmatrix}$ is orthogonal? (See Theorem 2.)

2. What are the relations among a, b, c, d such that $\begin{pmatrix} a & b \\ c & d \end{pmatrix}$ is the matrix of a magnification about $(0, 0)$?

3. Among the following matrices, decide which represent orthogonal transformations, or magnifications about $(0, 0)$, or neither.

a) $\begin{pmatrix} \frac{1}{2} & -\frac{1}{2} \\ \frac{1}{2} & \frac{1}{2} \end{pmatrix}$
b) $\begin{pmatrix} -2 & 0 \\ 0 & -2 \end{pmatrix}$
c) $\begin{pmatrix} -1 & 0 \\ 0 & 1 \end{pmatrix}$
d) $\begin{pmatrix} -2 & 0 \\ 0 & 2 \end{pmatrix}$
e) $\begin{pmatrix} .6 & .8 \\ .8 & -.6 \end{pmatrix}$
f) $\begin{pmatrix} 1 & 0 \\ 1 & -1 \end{pmatrix}$
g) $\begin{pmatrix} \frac{5}{13} & \frac{12}{13} \\ -\frac{12}{13} & \frac{5}{13} \end{pmatrix}$

4. For each of the linear functions whose matrices are given in Exercise 3, find the image of the point $(2, {}^-3)$. For example, the function of (a) maps $(2, {}^-3)$ onto $(\frac{5}{2}, -\frac{1}{2})$, since

$$\begin{pmatrix} \frac{1}{2} & -\frac{1}{2} \\ \frac{1}{2} & \frac{1}{2} \end{pmatrix} \begin{pmatrix} 2 \\ -3 \end{pmatrix} = \begin{pmatrix} 2 \cdot \frac{1}{2} - \frac{1}{2} \cdot {}^-3 \\ 2 \cdot \frac{1}{2} + \frac{1}{2} \cdot {}^-3 \end{pmatrix} = \begin{pmatrix} \frac{5}{2} \\ -\frac{1}{2} \end{pmatrix}.$$

5. Given the linear functions

$$(x, y) \xrightarrow{f} (2x - y, x), \qquad (x, y) \xrightarrow{g} (x + y, x - 2y),$$
$$(x, y) \xrightarrow{h} (y, x - y).$$

a) Without using matrix representation, find the following composite functions:

i) $f \circ g$ ii) $g \circ h$ iii) $h \circ f$ iv) $g \circ f$

[For example,

$$\begin{aligned}(f \circ g)(x, y) &= f(g(x, y)) = f(x + y, x - 2y) \\ &= (2(x + y) - (x - 2y), (x + y)) \\ &= (x + 4y, x + y).\end{aligned}$$

Hence $(x, y) \xrightarrow{f \circ g} (x + 4y, x + y)$.]

b) Write the matrices of f, g, h, and with matrix multiplication verify the composite functions obtained in part (a).

6. a) Write the matrix for each of the eight isometries in Exercise 9 of Exercise Set 6–2.

b) Show that this set of eight matrices (representing the symmetries of a square) forms a group under matrix multiplication. (Write a multiplication table of the products of the matrices and from this table verify all the group properties except associativity, which was already established.)

7. For given a, b, c, d solve these equations for u and r:

$$au + br = 1, \qquad cu + dr = 0$$

(Multiply the first equation by ${}^-c$ and the second by a; then add and solve for r. Follow a similar scheme to solve for u.)

8. Let f and g be any two invertible functions from a set S to S. Show that $f \circ g$ is invertible. (Since f^{-1} and g^{-1} exist such that $f \circ f^{-1} = f^{-1} \circ f = I$ and $g \circ g^{-1} = g^{-1} \circ g = I$, show that

$$(f \circ g) \circ (g^{-1} \circ f^{-1}) = (g^{-1} \circ f^{-1}) \circ (f \circ g) = I.$$

Why does this prove that the inverse of $f \circ g$ is $g^{-1} \circ f^{-1}$?)

9. a) Show that the composite of two rotations about (0, 0) is a rotation about (0, 0). [Let

$$\begin{pmatrix} a & -c \\ c & a \end{pmatrix}, \quad a^2 + c^2 = 1, \quad \text{and} \quad \begin{pmatrix} e & -f \\ f & e \end{pmatrix}, \quad e^2 + f^2 = 1,$$

be the matrices of two rotations about (0, 0) and show that their product matrix again represents a rotation about (0, 0).]
b) What is the composite of two reflections in a line through (0, 0)? [Let

$$\begin{pmatrix} a & c \\ c & -a \end{pmatrix}, \quad a^2 + c^2 = 1, \quad \text{and} \quad \begin{pmatrix} e & f \\ f & -e \end{pmatrix}, \quad e^2 + f^2 = 1,$$

be the matrices of two reflections in lines through (0, 0). What does their product matrix represent?]
c) What is the composite of two magnifications about (0, 0)? (See Exercise 2.)

10. Expand the results of Exercises 9(a) and 9(b) to show that the composite of two orthogonal transformations is orthogonal.

11. a) Using matrices find the inverse of a rotation about (0, 0). Is it also a rotation about (0, 0)? (Apply Theorem 3 to the matrix

$$\begin{pmatrix} a & -c \\ c & a \end{pmatrix}, \quad a^2 + c^2 = 1.)$$

b) Using matrices, find the inverse of a reflection in a line through (0, 0). Is it also a reflection?
c) Using matrices, find the inverse of a magnification about (0, 0). Is it also a magnification?

12. Which of the linear functions whose matrices are given below are invertible?

a) $\begin{pmatrix} -2 & 1 \\ 1 & 2 \end{pmatrix}$ b) $\begin{pmatrix} -2 & 1 \\ 4 & -2 \end{pmatrix}$ c) $\begin{pmatrix} 0 & 1 \\ 1 & 1 \end{pmatrix}$ d) $\begin{pmatrix} 2 & 0 \\ -3 & 0 \end{pmatrix}$

13. For the functions in Exercise 12 that are not invertible, find the set of all image points and show that they lie on a line through (0, 0).

14. If A, B, are two-by-two matrices, show that

$$\det(AB) = (\det A)(\det B).$$

Then use this result to show directly that the composite of two reflections in lines through (0, 0) is a rotation about (0, 0).

6-5
isometries and similarities

Let us summarize our study of the group T of (invertible) transformations of $\mathbb{P}$.

Subgroups of T:

a) Group E of isometries (length-preserving)

   Subgroups of E

   1) Group O of orthogonal transformations [with $(0, 0)$ a fixed point]

   2) Group Y of translations (no fixed points)

b) Group D of dilations (parallelism-preserving)

   Subgroups of D

   1) Group M of magnifications [with $(0, 0)$ a fixed point]

   2) Group Y of translations

c) Group L of linear (invertible) transformations

   Subgroups of L

   1) Group O of orthogonal transformations (with determinant 1 or $^-1$)

   2) Group M of magnifications about $(0, 0)$ (with determinant t^2, where t is the magnification faction)

The outline of subgroups makes it clear how to describe any isometry as a composite of an orthogonal function and a translation. For example, the function n,

$$(x, y) \xrightarrow{n} (^-y + 1, x - 3)$$

is an isometry because it is the composite $t \circ r$ of

$$(x, y) \xrightarrow{r} (^-y, x), \quad \text{the rotation of } 90° \text{ about } (0, 0),$$

and

$$(x, y) \xrightarrow{t} (x + 1, y - 3), \quad \text{the translation } [1, ^-3].$$

We calculate the fixed points of n:

$$^-y + 1 = x \text{ and } x - 3 = y \quad \text{imply} \quad x = 2 \text{ and } y = ^-1.$$

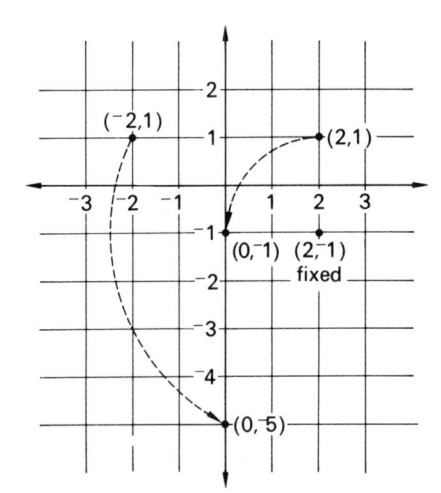

Figure 6-15

Since there is only one fixed point of n, namely $(2, {}^-1)$, then n is a rotation of $90°$ about $(2, {}^-1)$ (Fig. 6–15).

As another example, determine the isometry that reflects the plane in the line $x = y$ and then translates the plane by $[{}^-2, 3]$. Since the line $x = y$ is through $(0, 0)$ and is at a $45°$ angle with the positive x-axis, we consider the orthogonal reflection with $\theta = 2(45°)$ so that $\cos 90° = 0$, $\sin 90° = 1$, and the matrix of the reflection is $\begin{pmatrix} 0 & 1 \\ 1 & 0 \end{pmatrix}$. The composite of this orthogonal reflection and the translation $[{}^-2, 3]$ is

$$(x, y) \to (y - 2, x + 3).$$

Two figures α and β are congruent if there is an isometry of $\mathbb{P}$ that maps α onto β. Let us briefly consider functions of $\mathbb{P}$ that map figures onto *similar* figures. In loose language we say that a dilation of $\mathbb{P}$ stretches a figure uniformly in all directions from $(0, 0)$ and then translates the stretched figure, thereby preserving the shape but not the size of the figure. Line segments joining pairs of points of the figure are mapped by the dilation onto parallel segments k times as long. We say that the original figure is similar to the dilated figure. If we then move the dilated figure rigidly by an isometry, the resulting figure is still similar to the original figure. Thus we define a similarity as any transformation of $\mathbb{P}$ that is a dilation followed by an isometry. In these terms we say that figures α and β are similar if there is a function s in the set S of *similarities* of $\mathbb{P}$ such that s maps α onto β.

232   geometric transformations

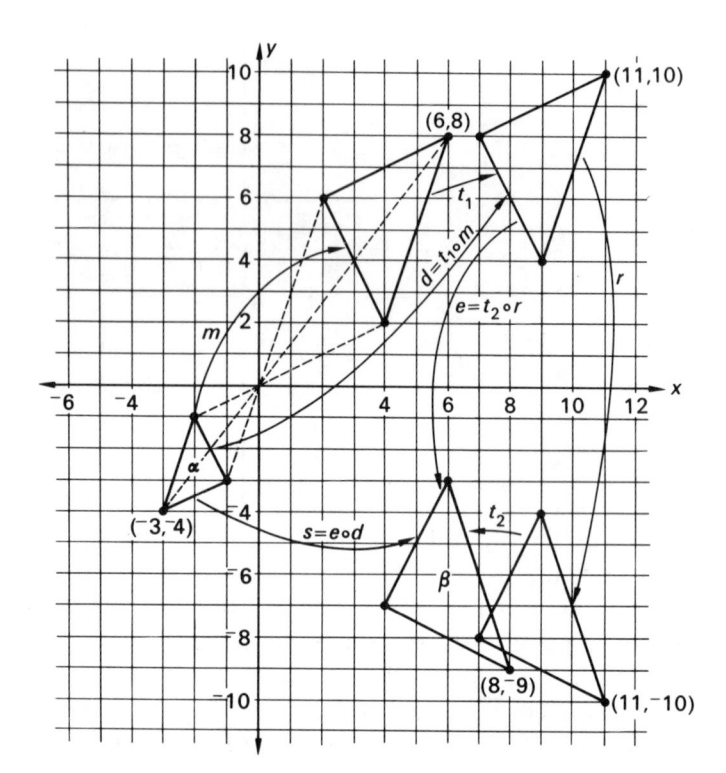

Figure 6-16

As an example, let α be the triangle with vertices $(^-2, ^-1)$, $(^-3, ^-4)$, $(^-1, ^-3)$ and let s be the similarity $e \circ d$, where d is the dilation $(x, y) \xrightarrow{d} (^-2x + 5, ^-2x + 2)$, and e is the isometry $(x, y) \xrightarrow{e} (x - 3, ^-y + 1)$. Note that d is $t_1 \circ m$, where m is the magnification $(x, y) \xrightarrow{m} (^-2x, ^-2y)$ and t_1 is the translation [5, 2]. Also e is $t_2 \circ r$, where r is the orthogonal transformation $(x, y) \xrightarrow{r} (x, ^-y)$ and t_2 is the translation $[^-3, 1]$. We may now regard s as the composite $t_2 \circ r \circ t_1 \circ m$ or the composite $e \circ d$:

$$(x, y) \xrightarrow{e \circ d} (^-2x + 2, 2y - 1). \quad \text{(Why?)}$$

In Fig. 6-16 the vertices of α are carried through the various mappings to the vertices of the image β. For example, let us follow the point $(^-3, ^-4)$:

$$(^-3, ^-4) \xrightarrow{s} (^-2(^-3) + 2, 2(^-4) - 1) = (8, ^-9)$$

and

$$(^-3, ^-4) \xrightarrow{m} (6, 8) \xrightarrow{t_1} (11, 10) \xrightarrow{r} (11, ^-10) \xrightarrow{t_2} (8, ^-9).$$

A more instructive way of writing this, so that the order of mappings is clearly indicated, is:

$$\begin{aligned}
s(^-3, ^-4) &= t_2 \circ r \circ t_1 \circ m(^-3, ^-4) \\
&= t_2 \circ r \circ t_1(6, 8), & \text{since } m(^-3, ^-4) &= (6, 8), \\
&= t_2 \circ r(11, 10), & \text{since } t_1(6, 8) &= (11, 10), \\
&= t_2(11, ^-10), & \text{since } r(11, 10) &= (11, ^-10), \\
&= (8, ^-9).
\end{aligned}$$

Would the function be changed by changing the order of any of its mappings in the composition? For example, is $s = t_2 \circ t_1 \circ r \circ m$? The reader should verify that this would be a different function from s, and that, in general, functions are not commutative under composition.

Finally, we remark without proof that the set S of similarities is a subgroup of the group T of (invertible) transformations of $\mathbb{P}$. To show this, we would need to prove (1) that the composite of two similarities is a similarity and (2) that every similarity is invertible. Note that (2) is easily proved by remarking that every similarity is a dilation d followed by an isometry e and that d and e are invertible. Hence, if $s = e \circ d$, then s^{-1} exists, and

$$s^{-1} = d^{-1} \circ e^{-1}.$$

(See Exercise 8, Exercise Set 6–3.)

Example. Show that the triangle A with vertices $(1, 2)$, $(3, 4)$, $(2, ^-1)$ is similar to the triangle B with vertices $(4, 0)$, $(8, 6)$, $(^-2, 2)$.

We must find a similarity s that maps A onto B. Assume that $(x, y) \xrightarrow{s} (ax + by + e, cx + dy + f)$. Then

$$\begin{aligned}
s(1, 2) &= (a + 2b + e, c + 2d + f) = (4, 0), \\
s(3, 4) &= (3a + 4b + e, 3c + 4d + f) = (8, 6), \\
s(2, ^-1) &= (2a - b + e, 2c - d + f) = (^-2, 2).
\end{aligned}$$

Equating components, we have two sets of equations to solve:

$$\begin{array}{ll}
a + 2b + e = 4 & c + 2d + f = 0, \\
3a + 4b + e = 8 & 3c + 4d + f = 6, \\
2a - b + e = ^-2 & 2c - d + f = 2.
\end{array}$$

These equations are true if

$$a = 0, b = 2, e = 0; \quad c = 2, d = 0, f = ^-2.$$

234 geometric transformations

Thus we have found a function mapping A onto B to be
$$(x, y) \xrightarrow{s} (2y, 2x - 2).$$
It remains to show that this function is a similarity. We can write
$$(x, y) \xrightarrow{m} (2x, 2y) \xrightarrow{r} (2y, 2x) \xrightarrow{t} (2y, 2x - 2),$$
so that $s = t \circ r \circ m$, where we recognize m as a magnification about $(0, 0)$, r as a reflection in the line $x = y$, and t as the translation $[0, {}^-2]$. Thus s is the magnification m followed by the isometry $t \circ r$, and hence it is a similarity. The reader may find it instructive to graph triangles A and B, and to trace the movements of A through the mappings of m, r, and t to B.

exercise set 6–4

1. Let $(x, y) \xrightarrow{r} ({}^-y, x)$, $(x, y) \xrightarrow{t} (x + 1, y - 3)$ be an orthogonal transformation and a translation. Find the composite $r \circ t$. Is it the same function as $t \circ r$, considered on page 230? If not, describe the function $r \circ t$ and calculate its fixed points.

2. For each of the following isometries, describe it as an orthogonal transformation followed by a translation; then calculate its fixed points and describe the type of isometry.
    a) $(x, y) \to ({}^-x + 2, y - 1)$
    b) $(x, y) \to ({}^-y - 3, x + 1)$
    c) $(x, y) \to (\frac{4}{5}x + \frac{3}{5}y - 1, \frac{3}{5}x - \frac{4}{5}y + 1)$
    d) $(x, y) \to (\frac{5}{13}x - \frac{12}{13}y + 2, \frac{12}{13}x - \frac{5}{13}y - 1)$
    e) $(x, y) \to ({}^-y, {}^-x - 3)$

3. Find each isometry from its description:
    a) A rotation of $45°$ about $(0, 0)$ followed by the translation $[{}^-2, 0]$.
    b) A reflection in the y-axis followed by the translation $[{}^-1, 2]$.
    *c) A rotation of $60°$ about $(0, 0)$ followed by the translation $[\sqrt{3}, 1]$. [Hint: $\cos 60° = \frac{1}{2}$, $\sin 60° = \frac{1}{2}\sqrt{3}$.]
    *d) A reflection in the line through $(0, 0)$ with inclination of $30°$ followed by the translation $[{}^-\frac{1}{2}, \frac{1}{2}]$.

4. a) In the example on page 232, verify that the composite $t_2 \circ r \circ t_1 \circ m$ is the function $(x, y) \to ({}^-2x + 2, 2y - 1)$.
    b) In the same example, find: (i) $t_2 \circ t_1 \circ r \circ m$; (ii) $t_2 \circ r \circ m \circ t_1$. Are they different from $t_2 \circ r \circ t_1 \circ m$?

5. Show that triangle A with vertices $(^-1, 1)$, $(^-1, 3)$, $(^-2, 1)$ is congruent to triangle B with vertices $(2, ^-2)$, $(4, ^-2)$, $(2, ^-3)$ by determining an isometry f that maps A onto B. (Assume that f is of the form $(x, y) \xrightarrow{f} (ax + by + e, cx + dy + f)$, calculate a, b, e, c, d, f such that f maps vertices of A onto vertices of B, and verify that the resulting function is an isometry.) Identify the isometry.

6. Show that triangle C with vertices $(1, ^-1)$, $(2, 0)$, $(^-1, 2)$ is similar to triangle D with vertices $(4, ^-5)$, $(1, ^-8)$, $(^-5, 1)$ by determining a similarity g that maps C onto D. Then identify the similarity as a certain dilation followed by a certain isometry. (Follow the example on pages 233–234.)

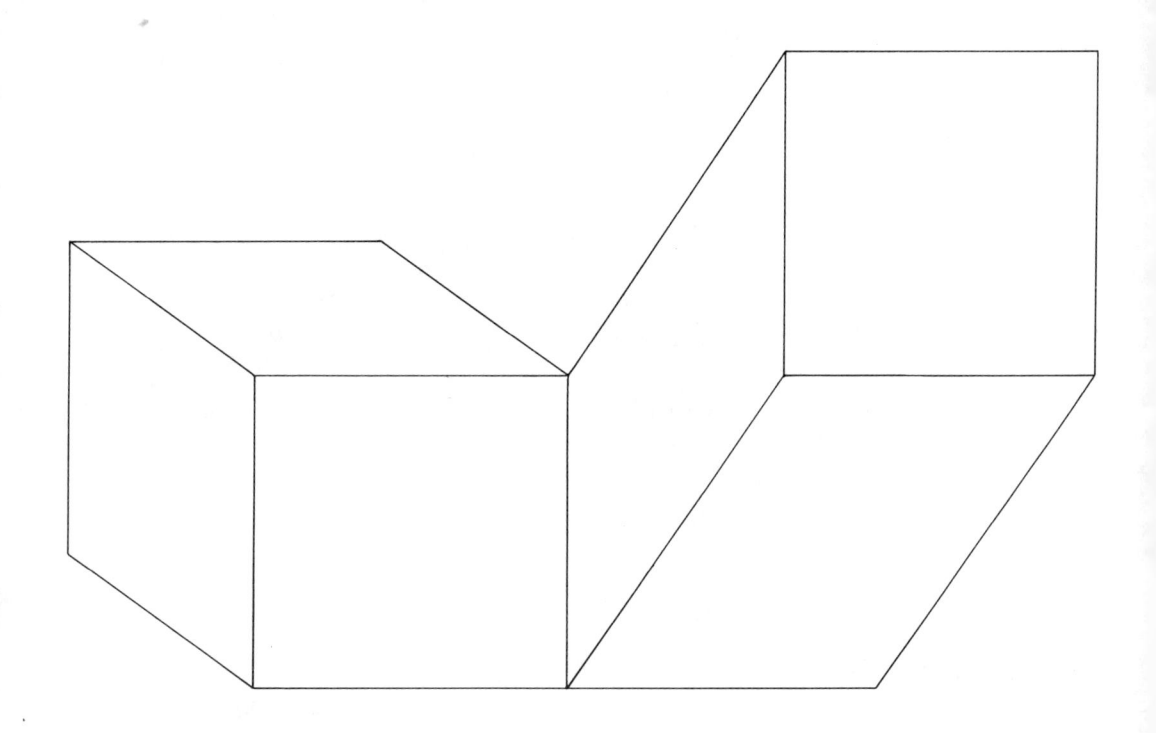

answers to selected exercises

exercise set 1–1

4. a) True   c) False   e) False
6. a) Simple closed curve   c) Closed curve   e) Curve
   g) Curve
8. $\overline{AB}$
10. a) *AEFB* and *DHGC* in prism; none in pyramid
    c) $\overleftrightarrow{AB}$ and $\overleftrightarrow{FB}$ in prism; $\overleftrightarrow{SP}$ and $\overleftrightarrow{RP}$ in pyramid
    e) *ABFE, ABCD, FBCG* in prism; *SPT, SPR, SRGT* in pyramid
    g) $\overleftrightarrow{AB}$ in prism; $\overleftrightarrow{SR}$ in pyramid

exercise set 1–2

2. b) Triangle
7. a) 4   b) 4
9. No
11. a) Infinitely many   b) Infinitely many
14. a) Parallel movement taking A to K
    c) Folding movement about a line taking C to C and D to L
15. a) Turning movement taking Flag l to Flag p, followed by parallel movement taking Flag p to Flag q.
17. No. A parallel movement along $\overleftrightarrow{FC}$ taking F to C (and G to G') followed by a folding movement about a line taking C to C' and G' to A.
19. a) Rhombus   b) Rectangle   c) Square

exercise set 2–1

1. a) 7   b) 6   c) 4   d) 3
3. c) 320   e) 91.44   g) 6080.2   i) 1760   k) 640
5. a) inch marks   c) .1-inch marks
7. a) $\frac{1}{2}$ inch   c) $\frac{1}{16}$ inch
9. a) 6 sq ft   b) 130 sq in.   c) 108 sq in.
12. a) Area is multiplied by 4   b) Area is divided by 4
13. a) 62.8 cm, 314.2 sq cm   b) 23.25 m, 43.01 sq m
    c) 32.7 in., 84.9 sq in.
15. a) Volume is doubled   c) Volume is multiplied by 4
    e) Volume is divided by 8
17. For $k = 8$, $\pi \approx 3.05$
18. a) 12π cu ft   c) 19.861π cu ft   e) $\frac{128}{3}$ cu ft   g) 72 cu in.

exercise set 2–2

3. a) $\frac{7}{4}$ is a rational number, hence a real number.
$$1 \leq \tfrac{7}{4} \leq 2$$
$$1.7 \leq \tfrac{7}{4} \leq 1.8$$
$$1.75 \leq \tfrac{7}{4} \leq 1.76$$
$$1.750 \leq \tfrac{7}{4} \leq 1.751 \quad \text{etc.}$$
$\frac{7}{4} = 1.750\ldots$

exercise set 3–1

1. a) The empty set   b) A set consisting of two points
   c) A set consisting of one point

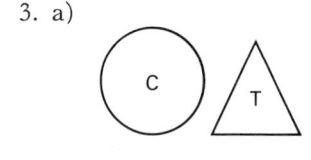

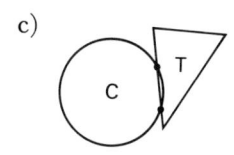

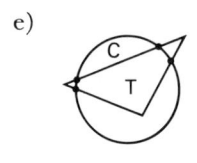

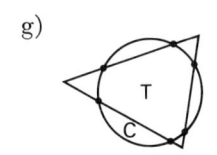

3. a)

   c)

   e)

   g)

4. a) $\overline{YU}$ c) $\overrightarrow{YV}$ e) $\overline{UV}$ g) $\overrightarrow{YV}$ i) $\overline{XU}$

9. Without the phrase "in a plane," Definition 5 would describe a sphere with center C.

13. If A, B, X are vertices of a triangle, then X does not belong to $\overline{AB}$, and, by Postulate 2(c), $AB < AX + XB$. This result holds for any relabeling of the vertices of $\triangle ABX$.

17. If lines l and k intersect, they intersect in one point, say Q. There is a point R in k different from Q, since there are at least two points belonging to each line (Postulate 1b). There is a plane $\mathbb{P}$ containing l and R, since R is not on l (Exercise 16). Plane $\mathbb{P}$ contains l and also contains k, since $\mathbb{P}$ contains two points R and Q of k (Postulate 1e). But no other plane contains l and k, since there is a point T in l different from Q and there is exactly one plane containing R, Q, and T (Postulate 1c).

18. b) 6 lines, 4 planes

20. b) Yes, since $\triangle PQR \equiv \triangle ABC$, by ASA (Postulate 5c), and $\angle PRQ$ and $\angle ACB$ are corresponding angles in these congruent triangles.

23. Outline of proof:
    $\overrightarrow{CP}$ is between $\overrightarrow{CA}$ and $\overrightarrow{CB}$, by definition of an angle bisector; $\overrightarrow{CP}$ intersects $\overline{AB}$ in a point D between A and B, by our assumption; $\triangle ACD \equiv \triangle BCD$, by SAS (Postulate 5b) because $\overline{AC} \equiv \overline{BC}$, $\angle ACP \equiv \angle BCP$, and $\overline{CD} \equiv \overline{CD}$. Then $\angle ADC \equiv \angle BDC$, since they are corresponding angles of congruent triangles, and $\angle ADC$ and $\angle BDC$ form a linear pair of angles. Hence $\angle ADC$ is a right angle, by Definition 7, and $\overleftrightarrow{AB} \perp \overleftrightarrow{CP}$, by Definition 9.

31. a) C_2 and C_3 intersect at two points, because $a = r$, $b = XY$, $c = r$, and each of these is less than the sum of the other two, by applying Postulate 2(c) to $\triangle XOY$ (or Exercise 13).
c) By SSS, since $\overline{OX} \equiv \overline{AE}$, $\overline{OY} \equiv \overline{AD}$ and $\overline{XY} \equiv \overline{ED}$.
e) $\angle XOY \equiv \angle EAD \equiv \angle FAD$

exercise set 3–2

3. a) $m(\angle y) = 50$, $m(\angle z) = 65$, $m(\angle x) = 65$
   b) $m(\angle y) = 40$, $m(\angle z) = 70$, $m(\angle x) = 70$

5. a) If a quadrilateral is a square, then the quadrilateral is a rectangle.
   c) If it is raining, then there are clouds in the sky.

6. a) If a quadrilateral is a rectangle, then the quadrilateral is a square. If there are clouds in the sky, then it is raining.
   b) None of the above is true.

7. a) If a quadrilateral is not a rectangle, then it is not a square. If there are no clouds in the sky, then it is not raining.
   b) Yes

11. If a triangle is isosceles, then two of its angles are congruent. This is a true statement; see Exercise 25, Exercise Set 3–1.

17. a) $\frac{22}{7} \approx 3.142857$, which agrees with the representation of π to two decimal places.
    b) $\pi - 3.14 \approx .00156$, $\frac{22}{7} - \pi \approx .00127$. Hence $\frac{22}{7}$ is a better approximation of π than 3.14.

exercise set 4–1

1. a) s is a function from A to B. $s(^-5) = 25$
   c) a is a function from A to B.
   e) f is a function from A to B. $f(^-3) = 6$
   g) q is a function from A to B. $q(^-\frac{3}{2}) = 6$
   i) d is not a function from A to B. $d(3, 6) = \frac{1}{2}$, not an integer
   k) c is a function from A to B. $c(2\frac{7}{8}) = 18$
   m) w is not a function from A to B. w does not assign to $^-4$ any real number, since there is no real number whose square is $^-4$.

2. a) Range of s is $\{0, 1, 4, 9, 16, \ldots\}$
   Range of a is set of all positive real numbers
   Range of f is set of all integers
   Range of q is set of all rational numbers
   Range of c is $\{6, 12, 18, 24, \ldots\}$
   b) Onto functions are f, q, m.

3. 1–1 functions are f, q.

5. g is a function from the integers to the rational numbers.

exercise set 4–2

1. Reflect one circle in k. If its image intersects the other circle, let the points of intersection be A and B. Reflect A in k to A'. Then $\overline{AA'}$ is bisected at C by k. Draw a circle with radius AC and center C, intersecting k at D and E. Then $ADA'E$ is a desired square.

3. Since $\overline{AB} \equiv \overline{CD}$, there is a reflection in a line l through B that takes A to C and C to A. Then $\overleftrightarrow{AC} \perp l$ and AC intersects l in a point D. Then the reflection takes $\triangle ABD$ to $\triangle CBD$, so that $\triangle ABD \equiv \triangle CBD$. Hence $\angle BAC \equiv \angle BCA$.

5. a) A 180° rotation about the center of the square
   b) A reflection in $\overleftrightarrow{AC}$

6. a) A translation, other than the identity translation, leaves no points of $\mathbb{P}$ fixed.
   c) A reflection leaves each point of a line in $\mathbb{P}$ fixed, and no other points.

exercise set 4–3

6. b) $r \circ t$ is a 60° clockwise rotation about O.

9. c) $\overleftrightarrow{P'P''} \parallel l$, by construction; $\overline{PA} \equiv \overline{P'A}$, $\overline{P'A} \equiv \overline{P''B}$, $\overline{P''B} \equiv \overline{DB}$, by construction; hence $\overline{DB} \equiv \overline{PA}$ and $\overleftrightarrow{PD} \parallel l$. But $\overleftrightarrow{DR} \parallel l$, by construction. Hence $\overleftrightarrow{PR} \parallel l$.

13. A rotation

15. Let $g_1 = t_1 \circ f$, where f is a reflection in line l and t_2 is the inverse of t_1. Then $g_2 = f \circ t_2$.

exercise set 4–4

1. a) Rhombus   c) Rectangle

3. a) Let f_1 be the reflection in one diagonal, f_2 the reflection in the other diagonal, r a 180° rotation about the center, and I the identity. The table of compositions is:

$\circ$	I	r	f_1	f_2
I	I	r	f_1	f_2
r	r	I	f_2	f_1
f_1	f_1	f_2	I	r
f_2	f_2	f_1	r	I

6. a) f_3 c) f_1 e) f_3

f) $\{(0, 0), (0, 1), (0, 2), (1, 0), (1, 1), (2, 0)\}$

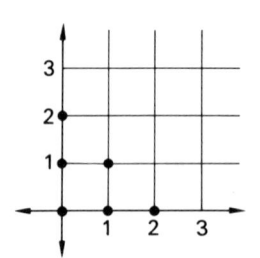

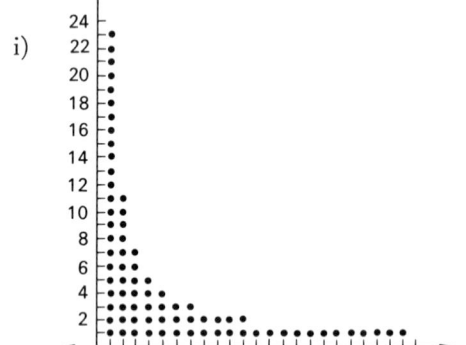

i)

Some ordered pairs:
$(1, 1), (1, 2), \ldots, (1, 23)$,
$(2, 1), (2, 2), \ldots, (2, 11)$,
$\ldots, (23, 1)$

5.

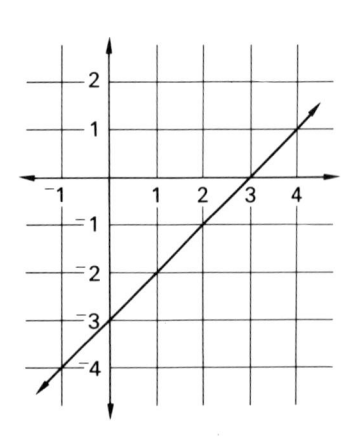

$f = \{(a, a - 3) \mid a \text{ is real}\}$
$= \{(x, y) \mid y = x - 3\}$,
x and y real numbers